Classical and Quantum Mechanics of Noncentral Potentials

A Survey of Two-Dimensional Systems

Classical and Quantum Mechanics of Noncentral Potentials

A Survey of Two-Dimensional Systems

R.S. Kaushal

Springer-Verlag

Narosa Publishing House

Dr. R.S. Kaushal
Department of Physics and Astrophysics
University of Delhi, Delhi-110 007, India

Exclusive distribution in Europe and North America (including Canada, Mexico and South America) by Springer-Verlag Berlin Heidelberg New York.

ISBN 3-540-64198-X Berlin Heidelberg New York
ISBN 81-7319-169-7 Narosa Publishing House, New Delhi

Printed in India.

To
My Revered Father
Late Sri Kshem Karan
Who Left a Number of Good Books at Home at the Time of his Departure for his Heavenly abode.

Nāsato vidyate bhāvo nābhāvo vidyate satah,
Ubhayorapi drastoantastvanayostatva darśibhih.
- Srìmad Bhagvad-Gìta (2/16)

The unreal has no existence, and the real never ceases to be; the reality of both (real and unreal) has thus been perceived by the seers of Truth.

Preface

In different branches of science and engineering there appear now several newly discovered phenomena whose theoretical understanding is definitely going to require an account not only of anharmonic but also of noncentral nature of the underlying forces. This book is an attempt to bring the available knowledge concerning some of the important classical and quantum aspects of such noncentral forces (potentials) in two dimensions at one place.

With a view to provide readymade results for immediate applications, time dependent and time independent systems are studied separately at the level of both classical and quantum mechanics. While some of the topics are discussed at a great length (such as the construction of invariants in the classical domain and the solution of the Schrodinger equation for such potentials in the quantum domain) others (such as Berry's phase, classical and quantum chaos, group- and field-theoretic aspects, extension of the results derived for 2D to the case of the corresponding 3D systems etc.) are given only a passing touch in the form of either a Section in the Chapter or an Appendix. I accept that by doing so I have not done justice to these topics which are growing rather exponentially these days but such limitations arose mainly by keeping the size of the book in mind and secondly, most of these topics are already discussed in other specialized books. This latter types of topics are, however, presented here only in an introductory style and that too for the sake of completeness. Again, as several reviews and books are now available on the classical mechanics of noncentral forces, particularly on the time independent case, the discussion of this aspect has been cut-short and relatively more emphasis has been placed on the quantum aspect and also on the classical aspect which pertains to the time dependent case. In the quantum context, though we investigate the solutions of the Schrodinger wave equation for noncentral potentials, but the methods presented are equally applicable to the case of all Schrodinger-like equations which frequently appear in different branches of mathematical sciences. The book can be useful not only to theoretical physicists and chemists, applied mathematicians but also to mechanical and electrical engineers.

For the past fifteen years and more the author has been actively engaged in pursuing research in some of the topics covered in this book. Several new and unpublished results of the author are also incorporated in addition to the published work of other authors. In any case, the author does not claim completeness of the survey of these ever growing topics covered in this book. In fact, some of them are still hot and timely from the point of view of further investigations.

In order to start any type of study of noncentral forces in two or higher dimensions, study of the corresponding (i) system in 1D, and/or (ii) central forces in the same dimensions, at least in terms of the mathematical techniques, is a prerequisite. It is only after the type (ii) studies that some meaningful (in the physical sense) results can be obtained which can highlight the importance of the underlying considerations of noncentral forces in describing the given physical phenomena. Therefore, in each Chapter, as and when required, a brief survey of the existing knowledge of the corresponding one dimensional (1D) system is also presented. The knowledge of 1D systems can also possibly offer some clue for developing the newer techniques for handling the noncentral systems in two and higher dimensions.

In this book, only the mathematical methods for handling the noncentral potentials at the level of both classical and quantum mechanics are presented. These methods, in fact, could be of immediate concern if one resorts to account for the noncentral character of the underlying forces of Nature, particularly for striking a better agreement between the theory and the experiment in various branches of physical sciences and engineering. Of course, I have nowhere in this book endeavoured for such a precise comparison between the theory and the experiment (and I do not think I am competent enough for this task) but the ideas presented will definitely make the theoretical front more sound against the experimental one which has gained a lot in recent years in terms of the advancing technologies.

The compilation of such a wide spectrum of ideas in the form of the present book is the first attempt of the author. Naturally, the human limitations are bound to reflect in different forms. I hope the readers will bear with me if they do not find the material and/or the citations in this book to their liking. Yet, I would very much appreciate any suggestion in this regard.

R.S. Kaushal

Acknowledgements

After devoting more than two years or so in writing the manuscript of this book, when I took the draft copy with me to the Fourth International Wigner Symposium held at Guadalajara (Mexico), several colleagues, whom I consider as experts in the subject, looked into the manuscript on my request and extended their suggestions and criticism in various forms. I have really benefitted from these suggestions and have incorporated many of them here. In this regard I am grateful to Professors V.I. Man'ko (Lebedav Inst., Russia), P. Winternitz (Univ. of Montreal, Canada) and S.S. Mizrahi (Univ. Federal de Sao, Carlos, Brasil). Immediately after that during my visit to the Univ. of Kaiserslautern (F.R.G.) Professors H.J. Korsch and H.J.W. Mueller-Kirsten went through the draft copy again and suggested several salient points. I am also thankful to them for sparing their valuable time for this purpose. I am sure some of the topics in the book could have never come to the present form without actually sharing the ideas with Prof. H.J. Korsch and with Prof. M. Znojil (Rez, Czech), in a truly friendly manner. The moral encouragement which I received from Prof. A.N. Mitra and Prof. R.P. Saxena (Univ. of Delhi) from time to time and from Prof. M. Moshinsky (UNAM, Mexico) during my visit to Mexico, deserve special gratitude. I also express my gratitude to Prof. N. Mukunda (I.I.Sc., Bangalore) for clarifying many ideas in the past on several occasions. The person to be acknowledged most and without whose help this project could have never been completed is my colleague, friend and collaborator Dr. D. Parashar (Univ. of Delhi). Thanks are also due to my senior student and colleague Dr. S.C. Mishra (Univ. of Kurukshetra) for reading parts of the manuscript critically and for several helpful suggestions.

An award of Research Scientist Scheme by University Grants Commission, New Delhi during the course of this work and a partial financial help from Av. H. Stiftung, Bonn during my visit to Univ. of Kaiserslautern, are gratefully acknowledged.

I am thankful to Mr. J.S. Patwal for his painstaking task in typing the manuscript in extra hours and Ms. Shraddha for her assistance in different form. Last but not the least important persons to be acknowledged are my wife Shashi and younger members of my family Shraddha, Medha, Mukta and Govind for the degree of the patience they have shown during the course of this project.

Finally, I thank M/s Narosa Publishing House, New Delhi, particularly Mr. M.S. Sejwal and Mr. N.K. Mehra, for the excellent job they have done in bringing the manuscript in the present form.

R.S. KAUSHAL

Contents

Classical and Quantum Mechanics of Noncentral Potentials

A Survey of Two-Dimensional Systems

1

General Introduction

1.1 NECESSITY FOR STUDYING NONCENTRAL FORCES

It is well known that in physics one often deals with the idealized models in which the central and/or harmonic nature of the underlying forces is frequently assumed mainly for the sake of mathematical simplicity. Although the predictive power of these models is sometimes overwhelming, but in a realistic situation an account of noncentral (NC) and/or anharmonic (AH) nature of the forces becomes necessary, particularly to bring the theoretical predictions and experimental results much closer. The effect of these latter type of forces is, in general, either neglected or computed in an approximate manner. It is partly due to the nonavailability of the exact formalism which can account for these forces and partly due to the fact that dealing with these forces as such turns out to be difficult at the conceptual level too in both classical and quantum mechanics. In fact, there exist now several newly discovered phenomena in various branches not only of physics and chemistry (such as structural phase transitions [1], polaron formation in solids [2], concept of false vacua in field theories [3], choice of models for various molecules [4], inhomogeneous waveguide problems in fibre optics [5], problems related to Aharonov-Bohm potential inside a thin infinitely long solenoid [6] etc.) but also of engineering whose theoretical understanding might require the introduction of higher order anharmonicities as well as an account of the NC nature of the potential function.

While the study of the noncentral anharmonic (NCAH) potentials has become much more desirable from the phenomenological point of view, it is equally important as a mathematical exercise in its own right, again at the level of both classical and quantum mechanics. Though in some cases the problem of AH oscillator potential (only with quartic-type anharmonicity) has been solved exactly [7] at the classical level in terms of elliptic functions, however, approximation methods are taken recourse to at the quantum level [8] for this purpose. Furthermore, in the latter case one either confines to one-dimensional problems or else only to central potentials in higher dimensions.

1.2 DIFFICULTIES IN DEALING WITH NONCENTRAL FORCES

As one starts working in two and higher dimensions, an increasing order of difficulties does arise in dealing with both harmonic and anharmonic NC forces. For example, at the classical level, not only the equations of motion turn out to be nonlinear in most cases, but also the existence of other dynamical invariants (which, if exist, can help in reducing the problem to quadrature) becomes questionable. At the quantum level, on the other hand, the Schrodinger equation for all the NC potentials although remains linear but there appear other types of difficulties concerning the energy-level spectra and the square integrability of the eigenfunctions.

A NC potential, in general, has an arbitrary dependence on the components of the position vector, but still a symmetrical occurence either of the components alongwith their powers in a functional form or of the powers of the components in a polynomial form often leads to some sort of simplification in their mathematical handling. Sometimes it is also possible to find a suitable transformation of the coordinates which can reduce the given NC potential either to a separable form or to a central one in the new representation. In the latter case while the problem reduces to a one-dimensional one, it can be solved by the method of separation of variables in the former case.

NC potentials can also be momentum- and/or angular momentum-dependent. Naturally, for such Hamiltonian systems momentum and/or angular momentum are not the constant of motion and one has to look for some other invariants. According to Whittaker [9], in the two-dimensional case one expects the existence of one more invariant besides the Hamiltonian. Sometimes this other invariant is called "second invariant" or "second constant of motion". No doubt the existence and subsequently the availability of the second invariant simplifies the solution of the problem but unfortunately it may not exist for all the NC potentials. Several methods have been developed in the literature for its construction. We shall return to these details in the next chapter.

For a particle moving in a NC potential $V(x_1, x_2)$ in the two-dimensional $x_1 x_2$-plane, the Lagrangian can be written as

$$L = \frac{1}{2}\left[\dot{x}_1^2 + \dot{x}_2^2\right] - V(x_1, x_2) \tag{1.1}$$

which gives rise to the following equations of motion

$$\ddot{x}_1 = -\frac{\partial V}{\partial x_1}, \qquad \ddot{x}_2 = -\frac{\partial V}{\partial x_2}. \tag{1.2a, b}$$

Note that for a harmonic potential of the type $V(x_1, x_2) = a_{20}x_1^2 + a_{02}x_2^2 + a_{11}x_1x_2$, equations of motion (1.2) although remain linear and coupled but they can admit a nontrivial solution for a number of choices of a_{20}, a_{02} and a_{11}. On the other hand, even for a simplest type of NCAH potential (say, for example, for $V(x_1, x_2) = a_{20}x_1^2 + a_{02}x_2^2 + a_{11}x_1x_2 + a_{30}x_1^3 + a_{03}x_2^3 + a_{21}x_1^2x_1 + a_{12}x_1x_2^2$) these equations of motion turn out to be nonlinear as well coupled. More often the solution of this latter class of problems turns out to be difficult with reference to the integrability of such systems. As mentioned above, in the quantum case however the nonlinearity problem does not arise. In fact, as long as the potential does not depend on the state-function, the Schrodinger equation remains linear for any form of $V(x_1, x_2)$. The square integrability of the eigenfunction sets a strong criterion for the existence of the solution. Such further details we postpone to Chapter 4. While the difficulties with the quantum mechanics of classically integrable systems can be handled to some extent, they are of higher [10] magnitude in the case of the nonintegrable systems.

1.3 CRITERION FOR NONSEPARABILITY OR NONCENTRALITY

As mentioned above, by using a suitable coordinate transformation the apparent NC nature of some of the potentials can be converted into a separable or into a central one. Although in this book, we shall investigate mainly the NC potentials but sometimes in order to demonstrate the elegance of the underlying mathematical technique, the cases of separable or that of central potentials of nontrivial nature will also be discussed. To ensure that the potential remains a noncentral one under a coordinate transformation, here, we analyse the classical and quantum cases separately. However, the conditions on the transformation matrix turn out to be identical in both the cases. Clearly, if $V(x_1, x_2) = f(\sqrt{x_1^2 + x_2^2})$, then the system has the radial symmetry.

Classical case: If $V(x_1, x_2)$ in (1.1) can be written as the sum of $u_1(x_1)$ and $u_2(x_2)$, then obviously the Lagrangian (1.1) is separable in x_1 and x_2 coordinates and this is not the case of interest for the present because the problem reduces to two separable one-dimensional motions. In general, if one performs a coordinate transformation

$$\xi = ax_1 + bx_2; \quad \eta = cx_1 + dx_2, \tag{1.3}$$

with $ad - bc \neq 0$, in the Langrangian (1.1) such that

$$V(x_1, x_2) = u_1(ax_1 + bx_2) + u_2(cx_1 + dx_2), \tag{1.4}$$

then L still remains separable and consequently the equations of motion remain uncoupled in ξ- and η-coordinates if and only if $ac + bd = 0$. In fact, such a condition makes the kinetic term in L separable in ξ and η. In other words, equations of motion (1.2) when expressed in ξ- and η-variables now get decoupled. Thus, for the nonseparability (noncentrality) one should have

$$ac + bd \neq 0. \tag{1.5}$$

Quantum case: In this case we look for the nonseparability of the Schrodinger equation

$$\left[\frac{\partial^2}{\partial x_1^2} + \frac{\partial^2}{\partial x_2^2} + \lambda - \upsilon(x_1, x_2)\right]\phi(x_1, x_2) = 0, \tag{1.6}$$

where $\lambda = 2\mu E/\hbar^2, \upsilon(x_1, x_2) = 2\mu V(x_1, x_2)/\hbar^2$, under the transformation (1.3). As for the classical case, if $\upsilon(x_1, x_2) = u_1(x_1) + u_2(x_2)$, then eq. (1.6) is also separable in the x_1- and x_2-coordinates; otherwise for the form (1.4) of $V(x_1, x_2)$, eq. (1.6) yields

$$\left[(a^2 + b^2)\frac{\partial^2}{\partial \xi^2} + (c^2 + d^2)\frac{\partial^2}{\partial \eta^2} + \lambda - u_1(\xi) - u_2(\eta)\right]\phi(\xi, \eta) = 0, \tag{1.7}$$

if and only if $ac + bd = 0$. Note that the operator in eq. (1.7) is separable in ξ- and η-coordinates and for its nonseparability, the condition (1.5) should hold. As such eq. (1.7) becomes solvable by the method of separation of variables.

An interesting situation arises when $V(x_1, x_2)$ can be written in the form

$$V(x_1, x_2) = u_1(x_1^2 + x_2^2) + u_2(x_2/x_1)/(x_1^2 + x_2^2).$$

In fact, the Schrodinger eq. (1.6) in this case becomes separable in the polar coordinates $r^2 = x_1^2 + x_2^2, \tan\theta = x_2/x_1$, as

$$\left[\frac{1}{r}\frac{\partial}{\partial r}\left(r\frac{\partial}{\partial r}\right) + \frac{1}{r^2}\frac{\partial^2}{\partial \theta^2} + \lambda - u_1(r) - \frac{u_2(\theta)}{r^2}\right]\phi(r, \theta) = 0,$$

and the corresponding classical equations of motion are not separable. Finally, note that if $V(x_1, x_2)$ is a homogeneous quadratic form in x_1 and

x_2, then there always exists an orthogonal transformation that diagonalizes $V(x_1, x_2)$ and leaves the form of the Laplacian invariant.

1.4 INTEGRABLE AND NONINTEGRABLE SYSTEMS

Integrability of a dynamical system is a concept which is very widely discussed for a long time now in both physics and mathematics and somewhat differently in different contexts. Here, however, we shall stick to the definition of integrability in the spirit of Liouville theorem [9]. According to Whittaker [9], if it becomes possible to obtain all the invariants or constants of motion of a dynamical system, then the system is said to be integrable, otherwise it is a nonintegrable system. Further, the number of functionally independent (cf. Sect. 1.6) invariants (in the sense of their pair-wise involuting property with respect to the Poisson bracket) which can exist for a system including the Hamiltonian itself is suggested to be the same as the dimensionality of the system. This is however true for a system which does not involve an explicit time-dependence (autonomous system). If a system involves explicit time-dependence (nonautonomous system), then the corresponding Hamiltonian also is not the constant of motion and one has to look for the other as many invariants as the dimensionality of the system. Sometimes the time is also considered as an additional dimension mainly for the purposes of having a unified mathematical treatment for both time-dependent (TD) and time independent (TID) systems (cf. Sect. 3.1). Here, however, we shall avoid such an abstraction or rather treat these cases separately so that the derived results could be of immediate use for the purposes of applying them to respective problems.

Since we plan to deal only with the two-dimensional systems, we shall be constructing the "second" invariant (besides the Hamiltonian) if it exists for the TID systems (cf. Chap. 2) and thereby confirming the integrability of the corresponding TID system. However, for TD systems it will not be possible to construct both the invariants in this book; therefore, the integrability of the corresponding two-dimensional TD systems is not ensured. Often we shall be constructing only one invariant and even this alone, if becomes available for a system, can simplify the understanding of the corresponding system considerably. It does not mean that the TD systems studied here are nonintegrable; rather it appears that the methods available for this purpose at present are inadequate to provide the second invariant for the two-dimensional TD systems. Although some general results are established [11] in this connection but they are valid only for a restricted class of dynamical systems.

It may be mentioned that if a system is integrable at the classical level it may not remain integrable at the quantum level. As a matter of fact in the classical limit ($\hbar \to 0$), a commutator does not reduce to a Poisson bracket as such but there arise some quantum corrections which can be ignored in the lowest order but not in higher orders of $\hbar$. These corrections however turn out to be important for some of the systems. We shall return to these details in Appendix 2.

Before closing the section a few remarks regarding the nature and the classification of the invariants are worth making: For the Hamiltonian systems, an invariant is basically a phase space function which, in general, can have any functional form. Though several mathematical forms have been investigated (cf. Chapter 2) in the literature for the TID systems and accordingly the invariants are constructed, but somehow the polynomial (in momenta or in velocities) form is found to be more suitable compared to the others for this purpose and for both TID as well as TD systems. Sometimes the order of the invariant is also designated by the degree of this polynomial. We refer to the work of Hietarinta [12] for the various ways of classifying the invariants. Most of the works on TD systems carried out thus far are focussed on the construction of the second order (in momenta) invariants only and for the TID systems however attempts have been made [13, 14] to construct third, fourth and higher order invariants in momenta. This is mainly because of the fact that TD systems do offer some additional difficulties in such constructions. While we postpone these latter details to Chap. 3, in Sect. 1.6 we shall make some formal remarks on the dynamical invariants.

1.5 DIFFERENT TYPES OF INVARIANTS

Like the concept of integrability, the notion of invariants is also very widely used in different disciplines of mathematical sciences. In this book, the word "invariant" or the "constant of motion" will be used with reference to the time evolution of the dynamical system. Even in this case various types of invariants are talked about in the literature. This is mainly because the construction of "exact" invariants of a given dynamical system has been a problem. In such a situation often "approximate" invariants either limited to subspace of the given phase space or to a limited time-dependence of the system, are designed. While Hall [14] constructed the "configurational" invariants for a variety of TID systems in two dimensions, the concept of "microscopic" or "adiabatic" invariants for TD systems has been prevailing [15] for a long time now.

The adiabatic invariants are defined normally for the systems having a mild dependence on time or are found as average over rapid nearly-periodic motions that take place within short times and within small spatial volume.

Adiabatic invariants can be used to convert the exact equations of motion into the approximate ones and thereby describing the behaviour in the large by appropriately glossing over behaviour in the small. On the other hand, configuration invariants are believed [14] to involve large scale motions and depend much more intimately on the nature of configuration itself. For a further explanation of the difference between the "exact" and "approximate" invariants at somewhat deeper level in terms of constant energy surfaces, we refer to the work of Hietarinta [12].

In fact, by noting that the Hamiltonian

$$H(x_1, x_2, p_1, p_2) = \frac{1}{2}\left(p_1^2 + p_2^2\right) + V(x_1, x_2), \tag{1.8}$$

while describes a mapping from the phase space to a real number system, the energy

$$E = \frac{1}{2}(p_1^2 + p_2^2) + V(x_1, x_2), \tag{1.9}$$

on the other hand, corresponds to a fixed number in this latter system and fixed by the initial conditions. Thus, H and E are the two conceptually different quantities in the sense that H maps the whole manifold whereas E corresponds only to its sub-manifold. As a matter of fact the Hamiltonian is a distinguished invariant because in testing an invariant one restricts to the manifold defined by the equation $H(x_i, p_i) - E = 0$, which implies the replacement of H by E. Accordingly the concept of integrability and subsequently that of exact and approximate invariants can be understood. In other words, eq. (1.9) sets a constraint on the phase space variables, thereby affecting the character of a dynamical invariant. Perhaps, the configuration invariants, as studied by Hall [14], are defined in this spirit. Some of these discussions will be continued in Chap. 8.

1.6 SOME FORMAL REMARKS ABOUT THE DYNAMICAL INVARIANTS

Although the theory of dynamical invariants is available (see, for example, Refs. [9] and [15-17]) now in a more rigorous mathematical language, but we restrict here only to some important formal results which will be of immediate use in the subsequent chapters. For the details, however, we refer to these classic works. Some of the ideas presented in the preceding section can, in fact, be put in this formal language as follows:

(i) Once the Hamiltonian, $H(x_i, p_i, t)$, of the system is known, then the time evolution of the coordinates and momenta is given by the Hamilton's equations of motion,

$$\frac{dx_i}{dt} = \frac{\partial H}{\partial p_i}; \qquad \frac{dp_i}{dt} = -\frac{\partial H}{\partial x_i}, \tag{1.10}$$

where $i = 1, 2$ is used all through this Section.

(ii) If $I(x_i, p_i)$ is another function in the given phase space, then its time evolution is given by

$$\frac{dI}{dt} = [I, H]_{\mathrm{PB}}, \tag{1.11}$$

where $[A, B]_{\mathrm{PB}}$ is the *Poisson bracket* defined as

$$[A, B]_{\mathrm{PB}} = \frac{\partial A}{\partial x_i}\frac{\partial B}{\partial p_i} - \frac{\partial A}{\partial p_i}\frac{\partial B}{\partial x_i}. \tag{1.12}$$

If I has to be a constant of motion (invariant), then

$$\frac{dI}{dt} = [I, H]_{\mathrm{PB}} = 0. \tag{1.13}$$

This implies that the constancy of I depends on the Hamiltonian, H, and in particular H itself is a constant of motion for TID (autonomous) systems and so are the functions of H. For the invariant I to be nontrivial, we must, therefore, require that I is *functionally independent* of H. Functional independence of two functions G and K can be tested by considering the 2×2 (or $2D \times 2D$ for the D-dimensional systems) Jacobian, $\partial(G, K)/\partial(x_i, p_i)$; if its rank is two, then G and K are functionally independent; otherwise they are said to be functionally dependent. For TD systems (nonautonomous), however, the time evolution of I is given by

$$\frac{dI}{dt} = \frac{\partial I}{\partial t} + [I, H]_{\mathrm{PB}}, \tag{1.14}$$

and again for the constancy of I one should have

$$\frac{dI}{dt} = 0, \tag{1.15}$$

as before.

In this book, while we will use the Hamiltonian approach with Poisson brackets as described above, the concept of integrability can also be formulated for the Lagrange method. In the latter case, the equations of motion follow from

$$\frac{d}{dt}\left(\frac{\partial L}{\partial \dot{x}_i}\right) - \frac{\partial L}{\partial x_i} = 0, \tag{1.16}$$

where $L = L(x_i, \dot{x}_i, t)$, is the Lagrangian of the systems.

(iii) A D-dimensional Hamiltonian system is said to be integrable (in the sense of Liouville [9]) or rather completely integrable if there exists a system of D functionally independent functions, I_n, with $n = 1, 2, \ldots D$, in the given phase space such that

$$[I_k, I_m]_{PB} = 0 \tag{1.17}$$

or $$[I_k, I_m]_{PB} = \sum_{i=1}^{D} \left(\frac{\partial I_k}{\partial x_i} \frac{\partial I_m}{\partial p_i} - \frac{\partial I_k}{\partial p_i} \frac{\partial I_m}{\partial x_i} \right) = 0$$

for all k and m, i.e. all I_n's are in involution. Note that for TID systems one of the I_n's is equal to H. In principle, there can be more than D functionally independent invariants, but then they can not be in involution. The maximum number of TID invariants is D, including the Hamiltonian, and if all of them exist and are globally defined and single-valued, then the system is sometimes called *superintegrable*. For further details about the invariants, we refer to the earlier works [12-17].

Besides the Langrangian and Hamiltonian formulations of classical mechanics, Hamilton-Jacobi theory also throws light on the dynamical invariants in terms of action- and angle-variables. However, we avoid these discussions here and refer the interested readers to the literature (see, for example, Ref. [16]).

1.7 SCOPE OF THE BOOK AND THE ARRANGEMENT OF THE CHAPTERS

The purpose of this book is to give some glimpses of the problems and their possible solutions pertaining to classical and quantum aspects of NC dynamical systems in two dimensions. As several recent reviews [12-14, 18] and also books [15-17] are now available on the classical aspect, the quantum aspect of these NC systems has not yet been explored to that extent. In spite of the fact that the study of classical aspect of NC systems is much richer and more transparent than the quantum one, in this book, however, our efforts will be to cover comparatively more on the quantum aspect. In the classical context, we shall restrict mainly to developing the methods for the construction of exact invariants for both TID and TD systems separately. On the other hand, in the quantum case we shall look for the exact or quasi-exact solutions of the Schrodinger equations for a variety of NC potentials. In most of the solvable cases one of the energy eigenvalues is obtained which often corresponds to the ground state. Prescriptions are also suggested to find the excited states. From the point of view of applications, even these ground state solutions can be used to develop the perturbation solutions for the related dynamical systems. Concerning these topics while the existing works are put in a coherent and systematic form, several new results also find a due place in different contexts, sometimes in the form of an unsolved exercise.

No doubt, some of the methods and techniques discussed here for 2D systems can easily be extended to the 3D case. However, they are not explicitly carried out in detail in this book except for making a few remarks in Chapter 6. This is mainly because it appears that without incorporating the knowledge of 3D systems (in whatever proportion it may be) the purpose of the monograph is defeated. Moreover not enough literature is available as far as the study of NC forces in 3D is concerned. Again full justice has not been done in this book with some of the topics which have developed as independent research subjects in recent years; for example, the topics of Berry's phase, classical and quantum chaos, constrained dynamics are discussed in Chapters 7 and 8 only in an introductory manner and that too in the context of NC forces in two dimensions. This is mainly necessitated by consideration of not to outgrow the size of the book. Furthermore, the discussion of invariants from the point of differential geometry or algebraic topology is completely avoided all throughout. Although the field and group theoretical discussions as such are avoided all through, yet a brief mention of some of these concepts is made in the end in the appropriate Appendices. For details of some of these aspects we refer to the book by Sudarshan and Mukunda [218].

From the title of the book one may get a feeling that there may be some discussion in the book on the quantum mechanics of NC potentials which appears in the context of two-body problems in nuclear and atomic physics. While this topic has already been dealt with in a great detail in various textbooks devoted to these subjects, any discussion on this aspect is knowingly avoided here. Contrary to this, some of the results obtained in this book (particularly in the context of momentum-dependent potentials) may find some applications in these branches of physics. The arrangement of the chapters is as follows:

In the next Chapter we present the classical aspect of NC TID systems. After introducing the various forms of the second invariant investigated in the literature, we continue discussing briefly the methods of construction of the second invariant. In particular, rationalization and Painleve methods are discussed. With a view to exploring some new integrable systems the complexification method is elaborated. In this and also in the subsequent Chapters we restrict ourselves to the study of polynomial forms of the invariants. Here, nonintegrable NC systems are also discussed in brief.

In Chapter 3, we concentrate on the classical aspect of TD systems. After making a few differentiating remarks on the exact and adiabatic invariants which are more relevant for the TD case, a survey of the well-studied one-dimensional TD systems is carried out in the introduction. Regarding the methods of construction even of the first invariant (as the Hamiltonian for TD systems is not the constant of motion), in this Chapter we present three methods. The first is the rationalization method as discussed in Chapter 2 with a possible extension to the TD case. The second is the Lie algebraic approach which is now applied to two-dimensional systems and the third is again the Painleve method. The problem of TD coupled oscillators is investigated with several possible generalizations. A new class of Ermakov-type systems in 2D is suggested. As the methods discussed in this Chapter are found to be inadequate from the point of view of providing the second invariant, the complete integrability of the studied NC TD systems could not be established.

Chapter 4 deals with the solution of the Schrodinger equation for a variety of NC TID potentials. In particular, an eigenfunction-ansatz method is used to study the quantum mechanics of NC AH potentials. The cases of cubic and quartic type anharmonicity are investigated. While the difficulties in dealing with the NC potentials are explicitly demonstrated, exact solutions of the Schrodinger equation are obtained for several solvable cases; but with certain constraints on the potential parameters. With the help of an example the role of such constraints is discussed. NC polynomial (up to quartic degree) and exponential potentials are investigated in detail. Several other (both exact and approximate) methods used in the literature to study the problem are also discussed here.

Quantum mechanics of NC TD systems is presented in Chapter 5. Here we attack the problem using two different approaches. The first one is a generalized version of the eigen-function-ansatz method of Chapter 4 which is used here for solving the TD Schrodinger equation for NC TD potentials. The second is the Feynman propagator approach of Lawande and his coworkers [19]. While the latter approach offers an alternative method and it is more convenient to apply it to some physical problems, the former one is used here to obtain the exact solution of the Schrodinger equation for a shifted TD harmonic oscillator in two dimensions.

In Chapter 6, we just make a few passing remarks on the applicability of the methods and the techniques developed and used to study the classical and quantum aspects of NC potentials in 2D in Chapters 2 to 5, to the case of NC systems in 3D. Although by using these methods and techniques of the 2D case some interesting and concrete results are discussed for the 3D case but more often as straightforward extensions. Again from the point of view of keeping the size of the book in mind, even such details are avoided. Thus, the remarks made in this Chapter essentially remain of qualitative nature.

Down-to-earth applications of the mathematical results derived in Chapters 2 to 5, are presented in Chapter 7. The role and the scope of the dynamical invariants is emphasized here in the context of various physical problems. It is true that it is not possible to attach a physical meaning to all the dynamical invariants constructed for a 1D or 2D system, nevertheless, as it is clear from this Chapter, their utility with reference to simplifying the solution of a number of physical problems cannot be ignored. After summarizing the possible interpretations attributed to these invariants in the literature, the scope and the role of these invariants are briefly discussed in the context of various branches of physics and mathematics.

From the point of view of further highlighting the role and scope of the invariants, studies of the dynamical systems in the presence of constraints are carried out in Chapter 8. In this Chapter we discuss briefly the role which constraints have to play in the construction of invariants in terms of Dirac brackets. Finally, we try to put our understanding achieved so far on the classical and quantum aspects of NC systems in a nut-shell in Chapter 9. Here, we also pose several problems in connection with the study of these NC systems and which still remain to be solved in future.

Finally, a word of caution about notations and abbreviations used in this book will be helpful at this very stage. While some abbreviations are used in localized manner, others are frequently used throughout including this Chapter. It is worth listing them here:

AH	≡ anharmonic
HO	≡ harmonic oscillator
NC	≡ noncentral
NCAH	≡ noncentral anharmonic
NCTD	≡ noncentral time dependent
NCTID	≡ noncentral time independent
nD	≡ n-dimension or n-dimensional ($n = 1, 2, 3$)
ODE	≡ ordinary differential equation(s)
PDE	≡ partial differential equation(s)
SE	≡ Schrodinger equation
TD	≡ time dependent
TID	≡ time independent
TDHO	≡ time dependent harmonic oscillator
TIDHO	≡ time independent harmonic oscillator
TDSE	≡ time dependent Schrodinger equation
TIDSE	≡ time independent Schrodinger equation

Unless it becomes essential to differentiate, we shall use the same symbol in this book for an operator (q-number) quantity and the corresponding c-number quantity. It is only the situation or the discussion which will make their meanings different. I hope that the reader will bear with me in this regard. Further, the word dimension with the system is used in the sense of its dependence on the space dimensions and the same is known as degrees of freedom of the system in literature. Hence there should not be any confusion.

2

Classical Mechanics of Noncentral Time Independent (TID) Systems

2.1 INTRODUCTION

An invariant of a system is basically a phase space function which is in involution with the Hamiltonian as well as with other invariants, if they exist for that system. Further as a physical requirement this function has to be a single-valued one. As mentioned in Sect. 1.4, for NCTID systems in two dimensions there may exist one more invariant besides the Hamiltonian. In the past, following the works of Darboux [21] and of Whittaker [9], several methods have been employed [12-14, 23] to construct this second invariant either in its exact or in approximate form. From the point of view of constructing this invariant, while not many attempts have been made to study the NCTID systems in polar coordinates, Cartesian coordinates are frequently used in the literature. As several reviews are available [12-14] on the use of Cartesian coordinates, in this as well as the next chapter relatively more emphasis will be on the use of complex coordinates $Z = x_1 + ix_2$, $\bar{Z} = x_1 - ix_2$, for this purpose. As a matter of fact this complexification method [23-26] is found not only to reproduce the known results but has also led [26, 27] to suggest several new integrable systems which perhaps could not be traced otherwise.

In the next section, we make a brief survey of various functional forms of the second invariant investigated in the literature and in Sect. 2.3, we continue with the construction of only the polynomial (in momenta) form of this invariant within the framework of the complexification method. In Sect. 2.4, we make a few observations on the TID nonintegrable systems i.e. on the systems for which the second invariant does not exist and hence cannot be constructed. Normally, such systems show some abnormal behaviour in terms of their phase-portraits. The other extreme of superintegrable systems is also discussed here in brief. Finally, the results are discussed and summarized in Sect. 2.5.

2.2 SEVERAL FORMS OF THE SECOND INVARIANT

No doubt the second invariant for a two dimensional TID system, in general, can be any arbitrary phase space function which is in involution with its Hamiltonian but, in the past, several specific forms of this invariant are investigated. For such a detailed survey we refer to the work of Hietarinta [12]. In what follows, we briefly mention some of these constructs mainly for the sake of completeness.

(a) Polynomial form of the invariant: For the systems not involving the momentum (or velocity) dependent forces (which normally is the case) the choice of the functional form of the second invariant is highly restricted. Particularly, in view of the fact that there is a special status of the kinetic term in the Hamiltonian, the second invariant is often found to possess a polynomial form in momenta. Further the order of the invariant is characterized by the degree of this polynomial form. While Whittaker's work was mainly restricted to the study of second order (quadratic) invariants, in later years, attempts have also been made [12-14] to study the third and higher order invariants. We shall return to some of these results in the next Section.

In general, for 2D systems an ansatz for the n-th order invariant can be made as

$$I_n = a_0 + \sum_{i_1=1}^{2} a_{i_1}\, \xi_{i_1} + \frac{1}{2!} \sum_{i_1, i_2=1}^{2} a_{i_1 i_2}\, \xi_{i_1}\, \xi_{i_2} + \ldots$$

$$+ \frac{1}{n!} \sum_{i_1, i_2 \ldots i_n=1}^{2} a_{i_1 i_2 i_3 \ldots i_n}\, \xi_{i_1}\, \xi_{i_2} \ldots\, \xi_{i_n}, \qquad (2.1)$$

where $\xi_i = \dot{x}_i$ and the coefficients $a_0, a_{i_1}, a_{i_1 i_2}, \ldots$ etc. are the functions of coordinates. Further, these coefficients are symmetric with respect to any interchange of their indices. Note that for the Hamiltonian (1.8) which is even power in momenta (i.e. it is invariant under the time-reflection symmetry), the form (2.1) will contain either even power or odd power terms in momenta (ξ_i). However, if the potential term in (1.8) involves momentum-dependence, then all the terms up to the desired order in (2.1) should be considered. Not only this, even the invariants with fractional powers in momenta (of course it will depend on the nature of momentum-dependence in the potential term of (1.8)) are permissible provided the physical situation allows the invariant to fulfil other mathematical requirements like single-valuedness or differentiability etc.

(*b*) *Rational form of the invariant:* Hietarinta [28] (for TID systems) and Lewis and Leach [29] (for TD systems) have considered the forms of the invariant which are the rational functions of p_1 and p_2. In particular, I is considered to have the form $I = R/S$, where R and S are polynomials in momenta and $S \neq 0$ so that not only S^{-1} exists but also $R\,S^{-1} = S^{-1}\,R$ holds. Clearly, the vanishing of the Poisson bracket of this form of I and the Hamiltonian (1.8) requires that $S[H,R]_{PB} = R[H,\,S]_{PB}$, which is equivalent to the pair $[H,\,R]_{PB} = GR;\ [H,S]_{PB} = GS$ with G as some rational function of momenta. While a further classification of the rational invariant is possible in terms of degrees of the polynomials R and S, a general result is established by Hietarinta [12]. According to him if $I = R/S$ is an invariant of the system then $I = (aR + bS)/(cR + dS)$, is also an invariant of the system provided the arbitrary constants a, b, c and d satisfy the condition $ad - bc \neq 0$. For further discussion on rational invariants we refer to the work of Hietarinta [12].

(*c*) *Transcendental form of the invariant:* A further generalized version of the rational invariants, called the transcendental invariants, is also investigated by Hietarinta [28]. By definition a transcendental invariant, I, is an arbitrary function K of two different polynomials R and S in momenta i.e.

$$I = K(R,\,S).$$

In this case, not only the functional form of K but also the degrees of polynomials R and S will suggest further classification of transcendental invariants. A typical integrable system admitting transcendental invariants is found to be [28] $H = \frac{1}{2}\,(p_1^2 + p_2^2) + x_2/x_1$. For further details on the transcendental invariants we again refer to the work of Hietarinta [12].

(*d*) *Other forms of the invariant:* From the point of view of further abstraction, one can also think of other possible generalized versions of invariants. For example, a product of a rational function and a transcendental function of polynomials in momenta can also represent the form of an invariant. In fact, the form of the invariant investigated by Hall [14], viz.

$$I = (P/Q)\exp(-R/S),$$

where P, Q, R and S are arbitrary polynomials in momenta with position-dependent coefficients, fall under this category.

It may be mentioned that the existence of an invariant of the type (b), (c) or (d) defined above will necessarily imply the existence of a simpler ((a)-type i.e. a polynomial form) invariant. Among other forms of the second invariant tried in the literature are the logarithmic functions [30] of momenta. In hydrodynamics and plasma physics several other forms of the invariant (depending upon a particular physical situation) also arise, but we avoid such details here. All through this book our emphasis will be on the construction and discussion of the polynomial form of the invariants only.

2.3 CONSTRUCTION OF THE SECOND INVARIANT: VARIOUS METHODS

In the past, several methods have been used for the construction of the second invariant such as the method of Whittaker [9] (termed here as the "rationalization" method); the Painlevé method, which is very widely used in recent years for the study of both TID and TD systems, and the Lax-Pair method. Here, however, we briefly emphasize the first two methods.

2.3.1 Rationalization Method

This method is again presented here in terms of the (a) Cartesian coordinates x_1 and x_2, and (b) complex coordinates $Z = x_1 + ix_2$ and $\overline{Z} = x_1 - ix_2$. No doubt, the Cartesian coordinates have very widely been used in most of the works including that of Whittaker but in recent years the complexification method has also led to several interesting results as far as the search for new integrable systems is concerned. Here, while we present the results only for the Cartesian case, the details are given for the complexification method.

(a) Case of Cartesian Coordinates

For the Lagrangian (1.1) and the concomitant equations of motion (1.2) we ask as to under what conditions on the coefficient functions $a_0, a_{i_1}, a_{i_1 i_2}, \ldots$ etc. in (2.1), polynomial form of the invariant (other than the total energy) could exist so that it is of second ($n = 2$), third ($n = 3$), fourth ($n = 4$), or higher order in momenta. For this purpose we consider the vanishing of the Poisson bracket of the Hamiltonian (1.8) and the form (2.1) of I_n through eq. (1.13). After using (1.2) for $\ddot{x}_1$ and $\ddot{x}_2$ and rationalizing the resultant expression with respect to the powers of $\dot{x}_1$, $\dot{x}_2$ and their all possible products, we arrive at a set of first order, coupled PDE's for a_0, $a_{i_1}, a_{i_1 i_2}, \ldots$ etc. The mutually consistent solutions of these coupled PDE's would finally converge into a PDE for V, called here the "potential" equation. The solution of this potential equation would directly provide the potential function admitting the invariant of desired order. For the simple case of linear invariants ($n = 1$) we refer to the earlier works [9, 12].

For the case of second order ($n = 2$) or quadratic invariants one arrives at the "potential" equation [9, 12, 13, 20]

$$3(c_1x_1 + c_2)(\partial V/\partial x_1) - 3(c_1x_1 + c_4)(\partial V/\partial x_2)$$
$$+ (c_1x_1x_2 + c_2x_1 + c_4x_2 - 2c_6)\left((\partial^2 V/dx_1^2) - (\partial^2 V/\partial x_2^2)\right) +$$
$$\left[c_1(x_2^2 - x_1^2) + 2c_2x_2 - 2c_4x_1 + 2c_3 - 2c_5\right](\partial^2 V/\partial x_1\partial x_2) = 0, \quad (2.2)$$

where c_i ($i = 1, \ldots, 6$) are the arbitrary constants of integration. This eq. (2.2), first obtained by Darboux [21], was later analysed by Whittaker [9] for a restricted choice of the constants c_i. Although this equation, while has been a subject of study to many authors [22] in recent years, we have investigated [20] two of its special solutions; namely, the separability of $V(x_1, x_2)$ in the x_1 and x_2 coordinates under both additive and multiplicative operations is considered.

As a result of a general analysis of eq. (2.2) several new integrable systems are obtained. For example, for the separable case $V(x_1, x_2) = f(x_1) + g(x_2)$, eq. (2.2) takes the form

$$3(c_1x_2 + c_2)(df/dx_1) - 3(c_1x_1 + c_4)(dg/dx_2)$$
$$+ (c_1x_1x_2 + c_2x_1 + c_4x_2 - 2c_6).\left[(d^2f/dx_1^2) - (d^2g/dx_2^2)\right] = 0. \quad (2.2')$$

Note that one of its solutions corresponds to

$$(df/dx_1) = c_1x_1 + c_4; \quad (dg/dx_2) = c_1x_2 + c_2,$$

which implies the integrable system,

$$V(x_1, x_2) = (1/2)\,c_1\,(x_1 + c_4/c_1)^2 + (1/2)\,c_1\,(x_2 + c_2/c_1)^2.$$

This is the case of a shifted harmonic oscillator in 2D for which the invariant can easily be derived [20]. Alternatively, eq. (2.2′) can also be written as

$$\left[(d^2f/dx_1^2) + (3(c_1x_2 + c_2)/(c_1x_1x_2 + c_2x_1 + c_4x_2 - 2c_6)).(df/dx_1)\right]$$
$$= \left[(d^2g/dx_2^2) + (3(c_1x_1 + c_4)/(c_1x_1x_2 + c_2x_1 + c_4x_2 - 2c_6)).(dg/dx_2)\right],$$

which for the cases (i) $c_2 = c_6 = 0$, and (ii) $c_4 = c_6 = 0$, becomes separable in x_1 and x_2 coordinates. Accordingly, the solutions of the resultant pair of ODE's for these cases will lead to the integrable systems. In fact, for the case (i), an integrable system [20] turns out to be

$$V(x_1, x_2) = (\lambda_1 / 8c_1^2)(c_1x_1 + c_4)^2 + (\lambda_1 / 8)x_2^2 - \left(f_0 / 2c_1(c_1x_1 + c_4)^2\right)$$
$$- \left(g_0 / 2x_2^2\right),$$

where f_0 and g_0 are integration constants and λ_1 is a separation constant. Similarly, the separation of $V(x_1, x_2)$ in the form $V(x_1, x_2) = f(x_1) . g(x_2)$ in eq. (2.2) also leads [20] to some interesting integrable systems.

Separability of $V(x_1, x_2)$ in some special forms like $(x_1^m + x_2^m)$ or $(x_1^m . x_2^n)$ is also analysed in the light of eq. (2.2). Several new integrable systems are derived and some of them are listed in Table 2.1 for the case when $V(x_1, x_2)$ is of the form $V(x_1, x_2) = x_1^m x_2^n$. Interestingly, the rationalization of (2.2) for this potential requires that $m + n = -2$ and $c_2 = c_4 = c_6 = 0$; $c_3 = c_5$. Consequently, the second invariants are constructed for a class of coupled oscillators even with the fractional powers of the coupling terms, but, of course, in a restricted domain of the x_1 x_2- plane. We shall return to some of these discussions in the next Chapter.

The case of third order ($n = 3$) or cubic invariant in momenta is also investigated by Holt [13]. For such cases of higher order it, however, becomes difficult to derive a single "potential" equation like (2.2) as for the second order case. In fact, for the third order case the over-determined system of couple PDEs satisfied by the coefficient functions a_{i_1}, $a_{i_1 i_2 i_3}$ leads to the following pair of potential equations and that too under a special condition:

$$Y\left(\partial^2 V / \partial x_1 \partial x_2\right) + (1/2)\, a_{222}\left(\partial V / \partial x_2\right) - \frac{1}{2} a_{111}\left(\partial V / \partial x_1\right)$$
$$= -\left(\partial^2 \phi(V) / \partial x_1 \partial x_2\right), \tag{2.3a}$$

$$Y\left(\partial^2 V / \partial x_2^2 - \partial^2 V / \partial x_1^2\right) - \frac{1}{2}\left(a_{111} + a_{222}\right)\left(\partial V / \partial x_2\right)$$
$$-\frac{1}{2}\left(a_{112} + a_{222}\right)\left(\partial V / \partial x_1\right)$$
$$= -\left(\partial^2 \phi(V) / \partial x_1^2\right) - \left(\partial^2 \phi(V) / \partial x_2^2\right), \tag{2.3b}$$

where ϕ is some function of V which implies the nonlinear nature of eqs. (2.3) in V and $Y = Z' - \phi(V)$, with Z' as a new function fixed by the following pair of equations

$$a_1 = Z'(\partial V/\partial x_1)\ ; \qquad a_2 = -Z'(\partial V/\partial x_2). \tag{2.4}$$

Here, again the expressions for a_i and $a_{i_1 i_2 i_3}$ involve the arbitrary constants of integration c_i's as before. Several cases corresponding to $\phi(V) = 0$, are studied by Holt [13]. The only known integrable case having $\phi(V) \neq 0$ is the one studied by Fokas and Lagerstrom [32] (cf. case (1) Table 2.2).

As far as the polynomial potentials in 2D are concerned a computer code has helped [31] in searching the integrable systems. Still analytical methods have to be used for a variety of potentials. The second invariant obtained for some important systems is listed in Table 2.2.

Table 2.1 Invariants for the case $m + n = -2$; $c_2 = c_4 = c_6 = 0$; $c_3 = c_5$ (cf. eq. (2.2))

S.No.	Values of m and n	Potential $V(x_1, x_2)$	Invariant (I)
1.	$m = n = -1$	$x_1^{-1}x_2^{-1}$	$(1/4)c_1\left\{2\left(x_1x_2^{-1} + x_2x_1^{-1}\right) + \left(\dot{x}_1x_2 - x_1\dot{x}_2\right)^2\right\}$ $+(1/2)c_3\left(2x_1^{-1}x_2^{-1} + \dot{x}_1^2 + \dot{x}_2^2\right)$
2.	$m = 1$, $n = -3$	$v_1x_1x_2^{-3}$	$(1/2)c_1v_1x_1x_2^{-3}\left(x_1^2 + x_2^2\right)$ $+(1/4)c_1\left(x_1\dot{x}_2 - \dot{x}_1x_2\right)^2$ $+(1/2)c_3\left[2v_1x_1x_2^{-3} + \dot{x}_1^2 + \dot{x}_2^2\right]$
3.	$m = -3$, $n = 1$	$v_2x_2x_1^{-3}$	$(1/2)c_1v_2x_2x_1^{-3}\left(x_1^2 + x_2^2\right)$ $+(1/4)c_1(x_1\dot{x}_2 - \dot{x}_1x_2)^2 +$ $(1/2)c_3\left[2v_2x_2x_1^{-3} + \dot{x}_1^2 + \dot{x}_2^2\right]$
4.	Superposition	$(v_1x_1x_2^{-3} +$ $v_2x_2x_1^{-3})$	$(1/2)c_1\left(v_1x_1x_2^{-3} + v_2x_2x_1^{-3}\right)\left(x_1^2 + x_2^2\right) +$ $(1/4)c_1\left(x_1\dot{x}_2 - \dot{x}_1x_2\right)^2 +$ $(1/2)c_3\left[2v_1x_1x_2^{-3} + 2v_2x_2x_1^{-3} + \dot{x}_1^2 + \dot{x}_2^2\right]$
5.	$m = -1/2$, $n = -3/2$ Superposition	$(v_1x_1^{-1/2}x_2^{-3/2}$ $+ v_2x_1^{-3/2}x_2^{-1/2})$	$(1/2)c_1\left(v_1x_1^{-1/2}x_2^{-3/2} + v_2x_1^{-3/2}x_2^{-1/2}\right)$ $\left(x_1^2 + x_2^2\right) + (1/4)c_1\left(x_1\dot{x}_2 - \dot{x}_1x_2\right)^2 +$ $(1/2)c_3\,[2v_1x_1^{-1/2}x_2^{-3/2} + 2v_2x_1^{-3/2}x_1^{-1/2}$ $+ \dot{x}_1^2 + \dot{x}_2^2]$

Table 2.2 Some important potentials in 2D admitting a polynomial form (n = 3) of the second invariant

Sr. No.	Potential $V(x_1, x_2)$	Invariant (I)	Ref. No.
1.	$(x_1^2 - x_2^2)^{-2/3}$	$(p_1^2 - p_2^2)(x_1p_2 - x_2p_1)$ $-4(x_2p_1 + x_1p_2)(x_1^2 - x_2^2)^{-2/3}$	Fokas et al [32], Inozemtsev [33]
2.	$((3/4)x_2^2 + x_1^2 + \delta)x_2^{-2/3}$	$2p_1^3 + 3p_1p_2^2 + 3p_1(-3x_2^{4/3}$ $+2x_1^2x_2^{-2/3} + 2\delta x_2^{-2/3}) + 18p_2x_1x_2^{1/3}$	Holt [13]
3.	$x_1^2 + 4x_2^2 + \delta x_1^{-2}$	$2p_1^2p_2 + 8x_1x_2p_1 + 2(-x_1^2 + \delta x_1^{-2})p_2$	Holt [13]
4.	$x_1^2/2 + x_2^2/18$	$(x_1p_2 - x_2p_1)p_2^2 + (1/27)x_2^3p_1$ $-(1/3)x_1x_2^3p_2$	Fokas et al [32]
5.	$\exp((\sqrt{3}x_1 - x_2)/2) +$ $\exp(x_2) +$ $\exp(-(\sqrt{3}x_1 + x_2)/2)$	$p_1^3 - 3p_1p_2^2 + 3\exp((\sqrt{3}x_1 - x_2)/2)$ $-2\exp(x_2) + \exp(-(\sqrt{3}x_1 + x_2)/2)p_1$ $+3\sqrt{3}[\exp((\sqrt{3}x_1 - x_2)/2)$ $-\exp(-(\sqrt{3}x_1 + x_2)/2)]p_2$	Toda [34], Hietarinta [12]
6.	$\lambda(x_1x_2)^{-2/3}$	$2\lambda(p_2x_2 - p_1x_1)(x_1x_2)^{-2/3}$ $+p_1p_2(p_1x_2 - p_2x_1)$	Inozemtsev [33], Kaushal et al [23]

The fourth and higher order invariants are also investigated but it really becomes difficult to handle the corresponding "potential" equations in general. Often some special solutions for V are obtained for highly restricted choices of the coefficient functions. We restrict ourselves from going into such details. However, a large number of cases studied according to the degree of the polynomial of the potential term can be found in earlier works [12].

(*b*) *Case of Complex Coordinates*

With a view to extending the list of integrable systems, complex coordinates $Z = x_1 + ix_2$ and $\bar{Z} = x_1 - ix_2$, are also used in place of x_1 and x_2 by Kaushal et al [23, 27] and Mishra [26]. In this case, we write the Lagrangian of the system as

$$L = \frac{1}{2}\,|\dot{Z}|^2 - V\left(Z,\overline{Z}\right), \quad \dot{Z} = \dot{x}_1 + i\dot{x}_2, \tag{2.5}$$

with the corresponding equations of motion

$$\ddot{Z} = -2\frac{dV}{d\overline{Z}}, \quad \ddot{\overline{Z}} = -2\frac{dV}{dZ}, \tag{2.6}$$

and make the polynomial ansatz (2.1) for the second invariant. However, now $\xi_1 = \dot{Z}$, $\xi_2 = \dot{\overline{Z}}$ and the coefficient functions $a_0, a_{i_1}, a_{i_1 i_2}$ etc. are the functions of Z and $\overline{Z}$ only. Upto fourth order we rewrite (2.1) in terms of some simplified notations as follows:

$$I = a_0 + a_i\xi_i + \frac{1}{2}a_{ij}\xi_i\xi_j + \frac{1}{6}a_{ijk}\xi_i\xi_j\xi_k + \frac{1}{24}a_{ijkl}\xi_i\xi_j\xi_k\xi_l, \tag{2.7}$$

where the summation over repeated indices is understood and $i, j, k, l = 1, 2$. Using $dI/dt = 0$, this form yields

$$a_{0,i}\xi_i + a_{i,j}\xi_i\xi_j + a_i\dot{\xi}_i + \frac{1}{2}a_{ij,k}\,\xi_i\xi_j\xi_k + \frac{1}{2}a_{ij}\left(\dot{\xi}_i\xi_j + \xi_i\dot{\xi}_j\right)$$

$$+\frac{1}{6}a_{ijk,l}\,\xi_i\xi_j\xi_k\xi_l + \frac{1}{6}a_{ijk}\left(\dot{\xi}_i\xi_j\xi_k + \xi_i\dot{\xi}_j\xi_k + \xi_i\xi_j\dot{\xi}_k\right)$$

$$+\frac{1}{24}a_{ijkl,m}\,\xi_i\xi_j\xi_k\xi_l\xi_m + \frac{1}{24}a_{ijkl}\,(\dot{\xi}_i\xi_j\xi_k\xi_l + \xi_i\dot{\xi}_j\xi_k\xi_l$$

$$+ \xi_i\xi_j\dot{\xi}_k\xi_l + \xi_i\xi_j\xi_k\dot{\xi}_l) = 0, \tag{2.8}$$

where the subscript after the comma to a's corresponds to their partial derivatives. Now, after accounting for the proper symmetrization of the coefficient functions and noting that (2.8) must hold identically in ξ's one obtains [23] the following conditions on a's

$$a_{ijkl,m} + a_{jklm,i} + a_{klmi,j} + a_{lmij,k} + a_{mijk,l} = 0, \tag{2.9}$$

$$a_{ijk,l} + a_{jkl,i} + a_{kli,j} + a_{lij,k} = 0, \tag{2.10}$$

$$a_{ij,k} + a_{jk,i} + a_{ki,j} + a_{ijkl}\,\dot{\xi}_l = 0, \tag{2.11}$$

$$a_{i,j} + a_{j,i} + a_{ijk}\,\dot{\xi}_k = 0, \tag{2.12}$$

$$a_{0,i} + a_{ij}\,\dot{\xi}_j = 0, \tag{2.13}$$

$$a_i\,\dot{\xi}_i = 0. \tag{2.14}$$

Eqs. (2.10), (2.12) and (2.14) yield the following set of PDEs:

$$(\partial a_{111}/\partial Z) = 0;\ \partial a_{222}/\partial \bar{Z} = 0,\ (\partial a_{122}/\partial Z) + (\partial a_{112}/\partial \bar{Z}) = 0, \quad (2.15a, b, c)$$

$$(\partial a_{111}/\partial \bar{Z}) + 3(\partial a_{112}/\partial Z) = 0;\ (\partial a_{222}/\partial Z) + 3(\partial a_{122}/\partial \bar{Z}) = 0, \quad (2.15d, e)$$

$$(\partial a_1/\partial Z) = a_{111}(\partial V/\partial \bar{Z}) + a_{112}(\partial V/\partial Z), \quad (2.15f)$$

$$\partial a_2/\partial \bar{Z} = a_{122}(\partial V/\partial \bar{Z}) + a_{222}(\partial V/\partial Z), \quad (2.15g)$$

$$(\partial a_1/\partial \bar{Z}) + (\partial a_2/\partial Z) = 2a_{112}(\partial V/\partial \bar{Z}) + 2a_{122}(\partial V/\partial Z), \quad (2.15h)$$

$$a_1(\partial V/\partial \bar{Z}) + a_2(\partial V/\partial Z) = 0, \quad (2.15i)$$

whereas eqs. (2.9), (2.11) and (2.13) yield

$$(\partial a_{1111}/\partial Z) = 0;\ (\partial a_{1111}/\partial \bar{Z}) + 4(\partial a_{1112}/\partial Z) = 0, \quad (2.16a, b)$$

$$2(\partial a_{1112}/\partial \bar{Z}) + 3(\partial a_{1122}/\partial Z) = 0;\ 3(\partial a_{1122}/\partial \bar{Z}) + 2(\partial a_{1222}/\partial Z) = 0, \quad (2.16c,d)$$

$$(\partial a_{2222}/\partial Z) + 4(\partial a_{1222}/\partial \bar{Z}) = 0;\ (\partial a_{2222}/\partial \bar{Z}) = 0, \quad (2.16e,f),$$

$$3(\partial a_{11}/\partial Z) = 2a_{1111}(\partial V/\partial \bar{Z}) + 2a_{1112}(\partial V/\partial Z), \quad (2.16g)$$

$$3(\partial a_{22}/\partial \bar{Z}) = 2a_{2222}(\partial V/\partial Z) + 2a_{1222}(\partial V/\partial \bar{Z}) \quad (2.16h)$$

$$(\partial a_{11}/\partial \bar{Z}) + 2(\partial a_{12}/\partial Z) = 2a_{1112}(\partial V/\partial \bar{Z}) + 2a_{1122}(\partial V/\partial Z) \quad (2.16i)$$

$$(\partial a_{22}/\partial Z) + 2(\partial a_{12}/\partial \bar{Z}) = 2a_{1222}(\partial V/\partial Z) + 2a_{1122}(\partial V/\partial \bar{Z}) \quad (2.16j)$$

$$(\partial a_0/\partial Z) = 2a_{11}(\partial V/\partial \bar{Z}) + 2a_{12}(\partial V/\partial Z), \quad (2.16k)$$

$$(\partial a_0/\partial \bar{Z}) = 2a_{12}(\partial V/\partial \bar{Z}) + 2a_{22}(\partial V/\partial Z) \quad (2.16l)$$

Note that eqs. from (2.15a) to (2.15i) correspond to third order invariants and those from (2.16a) to (2.16l) correspond to fourth order invariants. The solutions of eqs. (2.15a) to (2.15e) can be obtained as [23]

$$a_{1111} = \frac{1}{6}C_3\bar{Z}^3 + C_4\bar{Z} + C_5\ ,\quad a_{222} = -\frac{1}{6}C_3Z^3 + C_6Z + C_7$$

$$a_{112} = -\frac{1}{6} C_3 Z\overline{Z}^2 - \frac{1}{3} C_4 \overline{Z} + C_2 \ ; \ a_{122} = \frac{1}{6} C_3 \overline{Z} Z^2 - \frac{1}{3} C_6 \overline{Z} + C_1 \tag{2.17}$$

whereas those of (2.16a) to (2.16f) turn out to be

$$a_{1111} = -\frac{1}{6} D_3 \overline{Z}^4 - 2D_4 \overline{Z}^2 - 4 D_5 \overline{Z} + D_8,$$

$$a_{2222} = -\frac{1}{6} D_3 Z^4 - 2D_6 Z^2 - 4 D_7 Z + D_9,$$

$$a_{1112} = \frac{1}{6} D_3 Z\overline{Z}^3 + D_4 Z\overline{Z} + D_5 Z + D_1,$$

$$a_{1222} = \frac{1}{6} D_3 \overline{Z} Z^3 + D_6 Z\overline{Z} + D_7 \overline{Z} + D_2,$$

$$a_{1122} = -\frac{1}{6} D_3 Z^2 \overline{Z}^2 - \frac{1}{3} D_4 Z^2 - \frac{1}{3} D_6 \overline{Z}^2 + D_{10}. \tag{2.18}$$

where C_i's amd D_i's are the arbitrary constants of integration.

In order to eliminate a_1 and a_2 from eqs. (2.15f) - (2.15h), we differentiate (2.15i) with respect to Z and use (2.15f) to obtain an expression for $(\partial^2 a_2 / \partial Z^2)$. Now, differentiate this expression with respect to $\overline{Z}$ and obtain $(\partial^3 a_2 / \partial \overline{Z} \partial Z^2)$ which is also equal to $(\partial^2 / \partial Z^2)(\partial a_2 / \partial \overline{Z})$. Subsequently, the use of (2.15h) for $(\partial a_2 / \partial \overline{Z})$ in this quantity and the rearrangement of terms in the resultant expression leads [23] to the following "potential" equation for the third order invariants:

$$\begin{aligned}
&\left[\left(\partial^2 a_{111} / \partial \overline{Z}^2\right) + \left(\partial^2 a_{122} / \partial Z^2\right) - 2\left(\partial^2 a_{112} / \partial Z \,.\, \partial \overline{Z}\right)\right]\left(\partial V / \partial \overline{Z}\right) \\
&+ 2\left[\left(\partial a_{111} / \partial \overline{Z}\right) - \left(\partial a_{112} / \partial Z\right)\right].\left(\partial^2 V / \partial \overline{Z}^2\right) + a_{111}\left(\partial^3 V / \partial \overline{Z}^3\right) \\
&+ [(\partial^2 a_{112} / \partial \overline{Z}^2) + (\partial^2 a_{222} / \partial Z^2) - 2(\partial^2 a_{112} / \partial Z \cdot \partial \overline{Z})] \cdot (\partial V / \partial Z) \\
&+ 2\left[(\partial a_{222} / \partial Z) - (\partial a_{122} / \partial \overline{Z})\right](\partial^2 V / \partial Z^2) + a_{222}(\partial^3 V / \partial Z^3) \\
&- a_{112} (\partial^3 V / \partial Z \cdot \partial \overline{Z}^2) - a_{122} (\partial^3 V / \partial \overline{Z} \cdot \partial Z^2) = 0 .
\end{aligned} \tag{2.19}$$

In the same way one can proceed to eliminate a_{11}, a_{12}, and a_{22}, from eqs. (2.16g) – (2.16j) and arrive [23] at the following "potential" equation for the fourth order invariants:

$$[(\partial^3 a_{1122} / \partial Z \cdot \partial \bar{Z}^2) - \frac{1}{3} (\partial^3 a_{1112} / \partial \bar{Z}^3)$$

$$- (\partial^3 a_{1222} / \partial \bar{Z} \cdot \partial Z^2) + \frac{1}{3} (\partial^3 a_{2222} / \partial Z^3)] (\partial V / \partial Z)$$

$$+ \left[(\partial^2 a_{1122} / \partial \bar{Z}^2) - (\partial^2 a_{1222} / \partial Z \cdot \partial \bar{Z}) + (\partial^2 a_{2222} / \partial Z^2)\right] (\partial^2 V / \partial Z^2) +$$

$$\left[(\partial a_{2222} / \partial Z) - (\partial a_{1222} / \partial \bar{Z})\right] (\partial^3 V / \partial Z^3) + \frac{1}{3} a_{2222} (\partial^4 V / \partial Z^4)$$

$$+ \left[(\partial a_{1122} / \partial \bar{Z}) - (\partial a_{1222} / \partial Z)\right] \cdot (\partial^3 V / \partial \bar{Z} \cdot \partial Z^2)$$

$$- \frac{2}{3} a_{1222} (\partial^4 V / \partial \bar{Z} \cdot \partial Z^3) + \frac{2}{3} a_{1112} (\partial^4 V / \partial Z \cdot \partial \bar{Z}^3)$$

$$+ \left[(\partial a_{1112} / \partial \bar{Z}) - (\partial a_{1122} / \partial Z)\right] (\partial^3 V / \partial Z \cdot \partial \bar{Z}^2) + \left[(\partial^3 a_{1112} / \partial Z \cdot \partial \bar{Z}^2)\right.$$

$$\left. + \frac{1}{3} (\partial^3 a_{1222} / \partial Z^3) - (\partial^3 a_{1122} / \partial \bar{Z} \cdot \partial Z^2) - \frac{1}{3} (\partial^3 a_{1111} / \partial \bar{Z}^3)\right] (\partial V / \partial \bar{Z})$$

$$+ \left[(\partial^2 a_{1112} / \partial Z \cdot \partial \bar{Z}) - (\partial^2 a_{1122} / \partial Z^2) - (\partial^2 a_{1111} / \partial \bar{Z}^2)\right] \left(\partial^2 V / \partial \bar{Z}^2\right)$$

$$+ \left[(\partial a_{1112} / \partial Z) - (\partial a_{1111} / \partial Z)\right] (\partial^3 V / \partial \bar{Z}^3) - \frac{1}{3} a_{1111} (\partial^4 V / \partial \bar{Z}^4) = 0 . \tag{2.20}$$

The coefficients a_{ijk} and a_{ijkl} appearing in eqs. (2.19) and (2.20) are defined in (2.17) and (2.18) Further note that the potential $V(Z, \bar{Z})$ obtained from (2.19) has to be in conformity with the remaining eq. (2.15i) and similarly, $V(Z, \bar{Z})$ obtained from (2.20) has to be in conformity with (2.16k) and (2.16l). As a matter of fact the solution of (2.19) or that of (2.20) for $V(Z, \bar{Z})$ as such is a difficult task, but one can use the following recipe for the construction of the invariant.

For a given form of $V(Z, \bar{Z})$ the unknown constants C_i's or D_i's can be determined by rationalizing the potential eq. (2.19) or (2.20). Subsequently, the determination of other coefficients a_i for (2.19) and a_0, a_{ij} for (2.20) from eqs. (2.15f) – (2.15i) and (2.16g) – (2.16l), respectively, lead to the final form of the invariant from (2.7).

Following Holt [13], here also for the third order case the 'potential' eq.

(2.19) splits into the following pair of equations

$$Y(\partial^2 V/\partial Z^2) - 3a_{112}(\partial V/\partial Z) - a_{111}(\partial V/\partial \bar{Z})$$
$$= -\partial(\phi\,\partial V/\partial Z)/\partial Z,$$
$$Y(\partial^2 V/\partial \bar{Z}^2) + 3a_{122}(\partial V/\partial \bar{Z}) + a_{222}(\partial V/\partial Z)$$
$$= -\partial(\phi\,\partial V/\partial \bar{Z})/\partial \bar{Z}. \qquad (2.21)$$

These equations are in accordance with eqs. (2.3) for the Cartesian case. This complexification method has been applied successfully to a large number of systems. While the results are reproduced for the known cases (like the cases (2), (3), (5) and (6) of Table 2.2), only some special solutions of eqs.(2.21) have led to several new integrable systems. In particular, from the analysis of the "potential" equation derived [27] for the second order invariants, namely

$$3(c_1 Z + c_4)(\partial V/\partial Z) + (c_1 Z^2 + 2c_4 Z + 2c_5)(\partial^2 V/\partial Z^2)$$
$$-3(c_1 \bar{Z} + c_2)(\partial V/\partial \bar{Z}) - (c_1 \bar{Z}^2 + 2c_2 \bar{Z} + 2c_3)(\partial^2 V/\partial \bar{Z}^2) = 0, \qquad (2.21')$$

where c_1 is the separation constant and other c_i's are the constants of integration, a new class of Toda-type potentials which admits the second order invariants is obtained. For further details and other applications of this method we refer to earlier works [23, 26, 27].

2.3.2 Painlevé Method

The efforts of some of the mathematicians of the last century have been revived recently concerning the study of dynamical systems. In particular, the existence of single-valued and analytic solutions for differential equations which subsequently led to the notion of "integrability in the complex plane", was studied by Painlevé [36] and others [37]. In fact, it is altogether a different notion from the one followed in the previous subsection for the Hamiltonian systems. As several [18, 38] reviews and books are now available on the use of Painlevé conjecture, we only mention here the main steps of working of this method in connection with the construction of invariants and subsequently regarding the integrability test of a dynamical system.

The Painlevé method is applicable to both ordinary and partial differential equations with reference to their singularity analysis in the complex plane of the independent variable. As a matter of fact the general solution of

ODE's (or PDE's) may cease to be analytic at certain points, called singularities. These singularities could be of different types such as pole, branch point- or essential-type. Further, these singular points may depend on the constants of integration (or ultimately on the initial conditions of the problem) as the solutions are functions of these integration constants. Also, these singular points may lie any where in the complex plane of the independent variable and in such a situation they are called movable singular points; otherwise if the singularities do not depend upon the integration constants then they are called fixed singular points.

The underlying idea of the Painlevé method is to identify and characterize the nature of the above mentioned singularities admitted by the general solution in the complex plane of the independent variable, and to find conditions under which the solution is meromorphic or related to meromorphism. In spite of the fact that there does not yet exist a direct connection between this Painleve′ property of a differential equation and the integrability of a dynamical system in rigorous mathematical terms, several authors [18, 38] have used this method for searching the new integrable systems. Perhaps for this reason only the Painlevé method has some limitations as far as the construction of exact invariants is concerned. For both Hamiltonian and non-Hamiltonian systems a considerable success is achicved by using this method, in general, however it gives some hints about the existence and the nature of the second invariant. For further details about this method in the context of noncentral potentials in 2D we refer to a recent [18] review by Lakshmanan and Sahadevan.

2.3.3 Other Approaches

Besides the above methods, several other approaches have also been followed in the literature for constructing the invariants in a rather indirect but more analytical manner. Here we briefly mention about (a) Lax-pair method, and (b) Lie symmetries approach. However, the application of these methods to higher-dimensional systems are highly limited at present.

(a) Lax-pair method

This method for solving the nonlinear differential equations due to P.D Lax [39] is essentially based on the following theorem:

Given the two operators $L(x_1, x_2, t)$ and $A\ (x_1, x_2, t)$, satisfying the operator equation $[L, A] = -\ \partial L/\partial t$, where [,] is the commutator, the eigenvalue λ of L such that $L\psi = \lambda\psi$, is independent of t if and only if the corresponding eigenfunction ψ evolves in t according to $A\psi = \partial\psi/\partial t$. Note that if $[L, A] = -\partial L/\partial t$ is equivalent to the original Hamilton's equations of motion, then the terms in the expansion of det $(L - H)$ involve invariants and the latter are in involution.

One of the interesting applications of this theorem in the context of invariants is for the three-particle system with nearest neighbour interaction. In this case, it is well known that the scaled and reduced version of the Hamiltonian for such a two-dimensional system is given by [12]

$$H = \frac{1}{2}(p_1^2 + p_2^2) + V(2x_2) + U\left(\sqrt{3}x_1 + x_2\right) + W\left(-\sqrt{3}x_1 + x_2\right), \tag{2.22}$$

where in the potential term, V, U and W are functions of their arguments. In the Lax-pair method the meaningful ansatz for the p-dependent part of the Lax matrix, L, is diag (p_1, p_2, p_3) and as a consequence the third invariant is det L (the first two being ½ tr (L^2) which is the Hamiltonian, and tr (L^4) which is the second invariant) has the form $I_3 = p_1\, p_2\, p_3 + \dots$. This suggests that the reduced system (2.22) has a second invariant of the form

$$I = p_1^3 - 3\, p_1\, p_2^2 + a_1\, p_1 + a_2\, p_2\,, \tag{2.23}$$

where a_1 and a_2 turn out to be the functions of V, U and W as

$$a_1 = 3\,[-2V + U + W]\ ;\ a_2 = -3\sqrt{3}\,[U - W]\,.$$

Also, the functions V, U and W must satisfy the functional equation

$$V'(W - U) + U'(-V + W) + W'(-U + V) = 0,$$

for the existence of the invariant (2.23). One immediate example of the above situation is that of the Toda-type potential (cf. case (5) in Table 2.2) which also admits a cubic invariant. The case when all the three V, U, and W have the same functional form, was investigated by Calogero [40] and Kulish [40] and some particular special cases were later studied by Moser [41] and also by Calogero [41]. While we restrain ourselves here from giving further details of this method, we refer to several reviews (see, for example, Steeb and Euler in ref. [38]) for this purpose. Note that this method offers somewhat deeper mathematical insight than the Painlevé method with regard to the construction of the second invariant.

(b) Lie symmetries approach

This method, essentially based on the invariance analysis of the differential equations, was originally advocated by Sophus Lie about a century ago and in recent years several of its generalizations have been carried out. In view of the fact that the underlying governing equations in the process of identifying an integrable system, are basically the differential equations (whether ordinary or partial), such an analysis has immensely helped in locating the invariants, particularly for the systems involving the nonlinear PDE's. While the historical contribution of E. Noether [42] to this approach was mainly concerned with the study of space-time symmetries and the related invariants, other advancements in this direction have been made in recent years with reference to the study of hidden or dynamical symmetries.

Lutzky [43] has shown that the integrals of motion of a finite dimensional Lagrangian system can be obtained from the infinitesimal symmetries. Leach [44] applied this method to the case of TD systems. Further, Sarlet et al, [45] investigated nonlinear ODE's in the context of SL(3,R) symmetry. Later, coupled quartic anharmonic oscillator and Henon-Heiles system are studied by Lakshmanan et al [46] within this framework. In a recent work by Ranada [47], this method has also been applied to investigate the integrability of two- and three-body systems with a modified version of Toda Lagrangian. While we postpone the mathematically abstract version of this method to Appendix 3, here we outline this method to the case of 2D systems.

For the system (1.1), the Lagrange's equations of motion (1.16) can be rewritten as

$$\ddot{x}_1 = \partial L / \partial x_1 \equiv \alpha_1 \ (x_1, x_2) \ ; \ \ddot{x}_2 = \partial L / \partial x_2 \equiv \alpha_2 (x_1, x_2). \tag{1.2'}$$

We consider the invariance of the system (1.2′) under a one-parameter ($\in$) group of transformations:

$$t \to T = T(t, x_1, x_2, \dot{x}_1, \dot{x}_2, \in)$$

$$x_i \to X_i = X_i (t, x_1, x_2, \dot{x}_1, \dot{x}_2, \in) \ , \ (i = 1, 2), \tag{2.24}$$

with the associated forms $\dot{X}_i = dX_i / dT$. Then the corresponding infinitesimal transformations (in the neighbourhood of $\in = 0$) are

$$t \to T = t + \in \xi \ (t, x_1, x_2, \dot{x}_1, \dot{x}_2) + O(\in^2),$$

$$x_i \to X_i = X_i + \in \eta_i \ (t, x_1, x_2, \dot{x}_1, \dot{x}_2) + O(\in^2), \tag{2.25}$$

$$\dot{x}_i \to \dot{X}_i = \dot{x}_i + \in (\dot{\eta}_i - \dot{\xi} \ \dot{x}_i),$$

where $\dot{\eta}_i = \Gamma \eta_i$, $(i = 1, 2)$ and $\dot{\xi} = \Gamma \ \xi$, and

$$\Gamma = \frac{\partial}{\partial t} + \frac{\partial}{\partial x_1}\dot{x}_1 + \frac{\partial}{\partial x_2}\dot{x}_2 + \alpha_1 \frac{\partial}{\partial \dot{x}_1} + \alpha_2 \frac{\partial}{\partial \dot{x}_2} . \tag{2.26}$$

Then the invariance conditions for the equations of motion (1.2′) follows as

$$\ddot{\eta}_i - \dot{x}_i \ddot{\xi} - 2\dot{\xi}\,\alpha_i = E\,(\alpha_i), \quad (i = 1,2)\,, \tag{2.27}$$

where

$$E = \xi \frac{\partial}{\partial t} + \eta_1 \frac{\partial}{\partial x_1} + \eta_2 \frac{\partial}{\partial x_2} + (\dot{\eta}_1 - \dot{\xi}\,\dot{x}_1)\frac{\partial}{\partial \dot{x}_1} + (\dot{\eta}_2 - \dot{\xi}\,\dot{x}_2)\frac{\partial}{\partial \dot{x}_2} . \tag{2.28}$$

Clearly, eqs. (2.27) form an incomplete system in η_1, η_2 and ξ. Therefore, in order to solve (2.27), one has to assume specific forms for η_1, η_2 and ξ. One trivial choice is $\eta_1 = \eta_2 = 0$, ξ = const. To determine the existence of other nontrivial infinitesimal symmetries, one assumes η_1, η_2 and ξ to be polynomials in $\dot{x}_1$ and $\dot{x}_2$ and then finds t-, x_1- and x_2-dependence consistently, viz.,

$$\xi = \sum_{i,j=0}^{N} a_{ij}\, \dot{x}_1^i \dot{x}_2^j \;;\; \eta_1 = \sum_{i,j=0}^{N} b_{ij}\, \dot{x}_1^i \dot{x}_2^j \;;\; \eta_2 \;;= \sum_{i,j=0}^{N} c_{ij}\, \dot{x}_1^i \dot{x}_2^j \,, \tag{2.29}$$

where a_{ij}, b_{ij} and c_{ij} are functions of x_1, x_2 and t.

Having obtained the explicit forms of the infinitesimal symmetries η_1, η_2 and ξ and using the Lagrangian L and the Noether's theorem, the associated conserved quantity, if it exists, can be derived as

$$I = (\xi\,\dot{x}_1 - \eta_1)\frac{\partial L}{\partial \dot{x}_1} + (\xi\,\dot{x}_2 - \eta_2)\frac{\partial L}{\partial \dot{x}_2} - \xi L + f, \tag{2.30}$$

where the function $f \equiv f(x_1, x_2)$, is to be determined from

$$E\{L\} - \dot{\xi} L = \frac{\partial f}{\partial x_1}\dot{x}_1 + \frac{\partial f}{\partial x_2}\dot{x}_2, \tag{2.31}$$

so that I is an invariant. For example, for the set $\eta_1 = \eta_2 = 0$ and ξ = const., the function f turns out to be $V(x_1, x_2)$, and $I = H$, the Hamiltonian. It may be noted [48] that not all the symmetries leave the Lagrangian invariant even though they leave the equations of motion invariant. For those sets of symmetries that leave the Lagrangian invariant, one can use the Noether's theorem. Sarlet and Cantrijn [49] have further studied the applicability of the Noether's theorem to investigate the dynamical symmetries. We shall address ourselves to some of these details in the next Chapter on TD systems.

2.4 NONINTEGRABLE AND SUPERINTEGRABLE TID SYSTEMS

So far we have discussed in this Chapter the systems for which the second invariant exists and the same can be constructed, thereby ensuring the integrability of the corresponding system. But as mentioned before, there can exist systems for which either no second invariant exists or there exist more than one second invariants in 2D which may not be in involution with themselves (cf. Sect. 1.6). Such extreme cases are the examples, respectively, of nonintegrable and superintegrable systems. In what follows, we furnish some such examples.

Sometimes even a small change in the parameters of the potential makes the system integrable or nonintegrable. Among a large number of nonintegrable systems in 2D, the most studied example is that of the *Henon-Heiles system* [50]

$$V(x_1, x_2) = \frac{1}{2}(x_1^2 + x_2^2) + x_1^2 x_2 - \frac{1}{3} x_2^3. \tag{2.32}$$

This is the classic example for which the second invariant does not exist and it has suggested a case for systematic studies of the nature of regular and chaotic motions. Another interesting system for which the nonintegrability is established by Grammaticos et al (see, Ref. [52]) is that of the Kaw and Sen potential (see Kaw et al in Ref. [50]),

$$V(x_1, x_2) = (x_1^2 + x_2^2)^{1/2} + k\, x_2,$$

which arises in a simplified description of nonlinear coupling between the electromagnetic wave and the electron plasma wave. It may be noted that even a small change in the numbers in various terms in (2.32) leads [14, 55] altogether to a completely different behaviour of the system as far as the studies of the corresponding phase-portraits are concerned. Interestingly, while the system (2.32) is nonintegrable its other varied forms [51] like $V = (1/2)\,(x_1^2 + x_2^2) + (1/3)\, x_2^3 + x_1^2 x_2$ (with invariant $I = p_1 p_2 + x_1 x_2 + (1/3)\, x_1^3 + x_1 x_2^2$) or for that matter $V = (1/2)\,(Ax_1^2 + Bx_2^2) + D(2x_2^3 + x_1^2 x_2)$ (with invariant $I = D(-x_2 p_1^2 + x_1 p_1 p_2) + x_1^2(D^2 x_2^2 + D^2 x_1^2/4 + DAx_2) + (A - B/4)\,(p_1^2 + Ax_1^2))$ are found to be integrable.

Sometimes by taking higher multiples of the order of the second invariant the corresponding potential attains a generalized version. As a simple example, the central potential $V = (x_1^2 + x_2^2)^2 + A(x_1^2 + x_2^2)$ admits a first order (linear in momentum) invariant $(x_2 p_1 - x_1 p_2)$ i.e., the angular momentum. However, by doubling its order i.e., the second order invariant,

$$I = (x_2 p_1 - x_1 p_2)^2 + (B - A)\,(p_1^2 + 2x_1^4 + 2x_1^2 x_2^2 + 2A\, x_1^2),$$

now corresponds to the potential $V = (x_1^2 + x_2^2)^2 + Ax_1^2 + Bx_2^2$, with A and B as arbitrary constants. This latter case is studied by Choodnovsky and Choodnovsky [52] and Grosse [52]. Another related and interesting potential studied by Wojciechowski [53] is

$$V = (x_1^2 + x_2^2)^2 + Ax_1^2 + Bx_2^2 + C\, x_1^{-2} + D\, x_2^{-2},$$

for which the second invariant turns out to be

$$I = (x_2 p_1 - x_1 p_2)^2 + 2C(x_2/x_1)^2 + 2D(x_1/x_2)^2$$
$$+ (B - A)\,(p_1^2 + 2x_1^4 + 2x_1^2 x_2^2 + 2A\, x_1^{-2} + 2C\, x_2^{-2}).$$

A system can also admit two or more than two independent (in the sense of involuting property) second invariants and thereby implying its superintegrability. For example, the potential $V = Ax_1^2 + Bx_2^2 + Cx_1^{-2}$ (cf. case (3) Table 2.2) admits one second-order and one third-order second invariant. Hence it is superintegrable. Several other potentials like [12, 32] $V = A(x_1^2 + x_2^2) + Cx_1^{-2} + Dx_2^{-2}$; $V = x_1^2/2 + x_2^2/18$, and $V = a/r + [b/(r + x_2) + c/(r - x_2)]/r$ with $r = \sqrt{(x_1^2 + x_2^2)}$, also fall under this category.

2.5 CONCLUDING DISCUSSION

With a view to constructing the second invariant for NC, TID systems in 2D, the main emphasis in this Chapter has been on discussing the various methods for this purpose. While the construction of the second invariant is finally carried out in a polynomial (in momenta) form, several other of its forms are also discussed. As a matter of fact the existence of a polynomial form is desirable from the physics point of view; however, the occurence of any other functional form of the invariant can also simplify the solution of the problem.

While only passing remarks are made for the Painlevé and Lax-pair methods with reference to the construction of the second invariant, the rationalization method on the lines of Whittaker, on the other hand, is discussed in great detail. Again in view of the availability of extensive literature on the use of Cartesian coordinates x_1 and x_2, relatively more emphasis is placed here on the use of complex coordinates Z and $\bar{Z}$. The "potential" equations corresponding to the third and the fourth order invaraiants are presented whose solutions would directly provide the integrable systems. Only some special solutions of these PDEs have led to some interesting results. As a matter of fact from the point of view of obtaining new integrable systems, no doubt the solutions of eq. (2.20) for the fourth order invariants have yet to be explored, but at the same time there is still more to be understood from the general solutions of eq. (2.19) or, for that matter, of eqs. (2.3), (2.21) and (2.21′).

As far as the relative merits of the methods presented here are concerned a few remarks are in order. The rationalization method, although simple and straightforward in 2D, becomes tedius in higher dimensions. Further, it seems to work well not only for the Hamiltonian systems but also for some limited functional forms as well as order of the invariant. On the other hand, the Painlevé method, still to be supported by mathematical rigour as far as the construction of exact invariants is concerned, however offers a quicker way to peep into the nature and, to some extent, into the form of the invariant. As a matter of fact in both these methods an ansatz for the invariant is a primary requirement and this is not so in the Lax-pair method outlined in Sect. 2.3.3. But somehow this latter method, though used very frequently for solving the nonlinear differential equations, has been tested only on a restricted class of potentials. Difficulties in dealing with these methods further crop up for the systems in higher dimensions where one expects the existence of other [54, 55] invariants besides the second one. In fact, as will be emphasized in Chapter 3 on TD systems, none of these methods is capable of providing the required number of invariants for a system of given dimensions if the latter happen to be more than two. Sometimes one has to depend on more than one method to obtain the complete set of invariants of a system which is necessary for ensuring the integrability of that system.

Integrable systems sometimes behave as nearly [56] integrable ones in the presence of perturbation but at the cost of the exact nature of invariants. During the last three decades or so there has been considerable interest in the study of such integrable systems. Whereas before perturbation the motion could be described as regular, after perturbation the phase space is sharply divided [57] into the regions of regular and irregular (chaotic) motions. Computer calculations have made it possible to reveal the complexity and the detailed structure of these regions. Such examples while are very common in a mathematics course, frequently occur in plasma physics, statistical mechanics and astronomy. We restrict ourselves from going into such details here as our motivation has been to deal only with the systems admitting exact invariants.

3

Classical Mechanics of Noncentral Time Dependent (TD) Systems

3.1 INTRODUCTION

No doubt, from the point of view of mathematical abstraction, a TD, n-dimensional Hamiltonian system can be replaced [58] by an $(n+1)$-dimensional Hamiltonian system in which time appears as a new canonical coordinate, but for the practical applications of the theory of TD dynamical systems a separate account of its time-variable is inevitable. This is what we wish to pursue in this Chapter.

In continuation of Sections 1.5 and 1.6, here in this Section we first add some further remarks to the concept of exact and approximate (adiabatic) invariants and then make a brief survey of one-dimensional TD systems. A summary of various methods used for the construction of exact invariants for 1D systems is presented in the next Section. However, the details of some of these methods are given in Sect. 3.3 in the context of 2D systems with particular reference to the construction of at least one second order invariant for these systems. Further attempts are also made to derive somewhat general results for the third and higher order invariants in Sect. 3.4. A generalization of Ermakov systems and a new class of Ermakov-type systems, based on the results of Sect. 3.3. are presented in Sect. 3.5. In particular, the problem of coupled, TD, anharmonic and anistropic oscillators in 2D is investigated in this section. In Sect. 3.6. we discuss the integrability of TD systems in 2D, of course without actually ensuring the same for these systems. Finally, concluding remarks are made in Sect. 3.7.

3.1.1 Exact and Adiabatic Invariants

The study of adiabatic invariants has received [54,57,59,60] considerable attention in the literature often in connection with the motion of charged particles in a particular electromagnetic field and also in cosmological problems (cf. Chap. 7). Until recently, with TD systems the concept of adiabatic invariants was attached more or less in an inseparable manner at the level of both classical and quantum mechanics. While at the quantum level such a confusion still persists, at the classical level, however, several methods have been developed in recent years which can throw light in-depth on the nature of invariants for these systems. As a matter of fact it has become possible now to construct exact invariants for a number of TD systems. In particular, a TD harmonic oscillator system has very widely been studied.

For a dynamical system involving slow variation with respect to time (or for other physical systems in which a physical quantity changes slowly from one state to another with respect to an independent variable), the adiabatic invariants are defined in analogy with the adiabatic process in thermodynamics. For instance, if λ is a TD parameter of the system, then by slow variation we mean that $T(d\lambda/dt) \ll \lambda$, where T is the period during which λ varies only slightly. In other words, the functional dependence of $\lambda(t)$ on t is bounded above by an exponential function. Such a system is not closed and hence the energy of the system is not conserved. For a TDHO, in whch $\omega(t)$ (angular frequencey) varies slowly with t, the adiabatic invariant turns [60] out to be $I = E/\omega$. For a detailed survey of adiabatic invariants we refer to the works of Chandrasekhar [54] and Whiteman [57]. In this book, however, we shall confine our discussions to exact invariants. Sometimes even for the systems admitting exact invariants, the presence of perturbation allows the construction of approximate invariants. This, in fact, helps in finding the solution of the problem, although in an approximate manner. In this case, however, the question of the degree to which a quantity appears to be a constant during the successive orders of the perturbation parameter, remains an interesting one.

Historically, Kolsrud [61] studied exact quantum dynamical solutions for a class of TDHO sytems by introducing a unitary time-displacement operator. Later, Kruskal [62] developed a general asymptotic theory of nearly-periodic classical systems and derived the invariant for the TDHO system. In fact, in order to see a connection between the exact and approximate invariants one considers the system,

$$H = (1/2\epsilon)\ [p^2 + \omega^2\ (t)\ x^2], \qquad (3.1)$$

for which there exists an invariant,

$$I = (1/2)\ [(x/\rho)^2 + ((\rho\dot{x} - \in \dot{\rho}\ x)^2], \tag{3.2}$$

where $\rho = \rho(t)$, satisfies an auxiliary equation

$$\in^2 \ddot{\rho} + \omega^2(t)\ \rho = \rho^{-3}. \tag{3.3}$$

Eqs. (3.2) and (3.3) define a class of invariants because ρ may be any particular solution of (3.3). Now, if (3.3) is solved recursively to give ρ as a series in positive powers of $\in$, then that value of ρ can be substituted into (3.2) to give I as a series in $\in$. For a classical system with real ω, that series for I is the usual adiabatic (approximate) invariant whose leading term is proportional to $\in H/\omega$. Kruskal's theory may be applied in a closed form to a system expressed in terms of x, $\dot{x}$ and $\rho(t)$ without actually demanding the adiabatic result in the limit of small $\in$. Further details of the Kruskal theory are left to the interested readers. Lewis [63] obtained an exact invariant for TDHO and studied the same in the context of both classical and quantum mechanics. For the quantum case, however, a derivation of simple relation between eigenstates of such an invariant and the solution of the Schrodinger equation has been studied by a number of authors [64]. It may be mentioned that the TD phase associated with the eigenstates of the invariant satisfies a simple first order nonlinear differential equation.

3.1.2 Survey of One-dimensional TD Systems

As mentioned before, the Hamiltonian of a system in 1D involving explicit time-dependence is not a constant of motion and one has to look for the other invariant of the system. The most studied case is that of a TDHO, described by the Hamiltonian (dropping $\in$ from eq. (3.1))

$$H = (1/2)\ [p^2 + \omega^2(t)\ x^2], \tag{3.4}$$

which admits the invariant,

$$I = (1/2)\ [k(x/\rho)^2 + (\rho\dot{x} - \dot{\rho}x)^2], \tag{3.5}$$

with $\rho(t)$ satisfying

$$\ddot{\rho} + \omega^2(t)\ \rho = k\ \rho^{-3}. \tag{3.6}$$

It may be mentioned that t-dependence, although shown in the potential term in (3.4), may arise through any one or both the terms in H. For example, in physical problems the pendulum with time varying mass and/or length can give rise [65] to such systems. Also, the damping (if present) can involve t-dependence. Note that the appearance of the constant k in (3.5) and (3.6) is illusory. While it may be important for physical reasons (cf. Chap. 7), it can as well be eliminated by a scale transformation $\rho \to \sqrt{k}\,\rho$. In any case, there exists a transformation (see, for example, Ref. [66]) which converts a TD damped system into a TD undamped one and subsequently a rescaling of x- and t-variables leads to the form (3.4). For example, the Lagrangian corresponding to the equation of motion (damped case).

$$\ddot{x} + f(t)\,\dot{x} + \omega^2(t)\,x = 0 \tag{3.7}$$

can be expressed as [67]

$$L = (1/2)\,e^{F(t)}\,[\dot{x}^2 - \omega^2(t)\,x^2], \tag{3.8}$$

where $dF/dt = f(t)$. If one defines the conjugate momentum $p = \dot{x}\,e^F$, then the corresponding Hamiltonian becomes $H = (1/2)\,[p^2\,e^{-F} + \omega^2(t)\,e^F\,x^2]$, describing a pendulum with time-dependence in both mass and frequency. In general, the system

$$\ddot{y} + f(t)\,\dot{y} + \overline{\omega}^2(t)\,y = G(t), \tag{3.9}$$

with arbitrary TD functions $f(t)$, $\overline{\omega}^2(t)$ and $G(t)$, can be cast in the form [68]

$$\ddot{x} + \omega^2(t)\,x = g(t) \tag{3.10}$$

by using the well known transformation

$$x = y\,\exp\,[(1/2)\textstyle\int^t f\,dt], \tag{3.11}$$

and with

$$\omega^2(t) = \overline{\omega}^2(t) - (1/2)\,\dot{f} - (1/4)\,f^2;\; g(t) = G(t)\,\exp\,[(1/2)\textstyle\int^t f\,dt].$$

The problem of TD anharmonic osciliator with cubic anharmonicity was investigated by Leach [69] and an exact invariant was obtained using the method of Lie theory of extended groups. The system he considered is,

$$\ddot{x} + a(t)\,\dot{x} + b(t)\,x + c(t)\,x^2 + d(t) = 0. \tag{3.12}$$

In a recent paper by Leach and Maharaj [70], the first invariant is constructed also for a class of TD, anharmonic oscillators of more complex type, namely the system,

$$\ddot{x} = a_1(t)\,x^2 + a_2(t)\,x^3 + a_3(t)\,x^4 + a_4(t)\,x^5,$$

has been studied. In fact, some particular cases of this equation arise in the study of charged plasma in an axially symmetric magnetic field and in a shear-free spherically symmetric gravitational field in general relativity.

Another TD anharmonic system which has been of interest [54, 71] is the modified Emden equation.

$$\ddot{q} + \alpha(t)\,\dot{q} + q^n = 0 \quad (n = \text{positive integer}),$$

which arises in the study of a spherical gas cloud acting under the central attractions of its molecules and subject to the laws of thermodynamics. The first invariant is constructed for this system by Leach [71] using a Lie point symmetry analysis.

The case of nonlinear equations of motion was also considered by Ray and Reid [72] using the Noether's theorem and by Kaushal and Korsch [66] using the dynamical algebraic approach, corresponding to the system

$$L = (1/2)[\dot{x}^2 - \omega^2(t)\,x^2 - \phi(x, t)]. \tag{3.13}$$

Interestingly, the system is found to admit an invariant for the case when $\phi\,(x, t) = 2F_0 G(t)\,x^{-2m}$, where m is an arbitrary constant, in the method of Ray and Reid. On the other hand, in the dynamical algebraic approach ϕ satisfies a PDE whose one of the particular solutions is the same as that obtained by Ray and Reid. The case when ϕ is momentum-dependent (instead of x-dependent) is also investigated in the dynamical algebraic approach and accordingly an invariant is constructed for the form $\phi\,(p, t) = 2G_0\,\eta\,(t)\,p^{-2m}$. Besides the above cases several generalizations of $\phi\,(x, t)$ in terms of the auxiliary variable $\rho(t)$ have also been considered in the literature. We shall return to some of them in the foregoing Sections.

3.2 METHODS FOR ONE-DIMENSIONAL TD SYSTEMS

There have been several methods developed in the past in order to obtain the invariant for TD systems in 1D. Sometimes the system (3.4) or its other related forms have offered the testing ground for deciding the merit of a method used for this purpose. While it may be more appropriate to discuss the details of these methods in the context of 2D systems in the next Section, here it is worth giving only a brief summary of them for the 1D systems. Besides the rationalization method of Whittaker-type, the other methods which we wish to emphasize below are Ermakov method, dynamical algebraic approach, Lutzky's approach using Noether's theorem (related to the Lie symmetries approach), transformation-group method and a few others.

3.2.1 Rationalization Method

Here, in analogy with eq. (2.1), one makes [73] an ansatz for the n-th order invariant as

$$I_n = b_0 + b_1\,\dot{x} + (1/2!)\,b_2\,\dot{x}^2 + \ldots\ldots + (1/n!)\,b_n\,\dot{x}^n\,, \tag{3.14}$$

where b_i are the coefficient functions, $b_i \equiv b_i\,(x, t)$. Note that unlike the TID case, here all powers in $\dot{x}$ appear in I_n up to a given n; this, in fact, complicates the applicability of the method for the TD case, particularly in higher dimensions, as will be clear from the next section. Now, for the Hamiltonian

$$H = (1/2)\,p^2 + V(x, t), \tag{3.15}$$

the use of eqs. (1.14) and (1.15) will yield a recursion relation for b_i's as

$$\dot{b}_i + i\,(\partial\,b_{i-1}/\partial x) - b_{i+1}(\partial V/\partial x) = 0, \tag{3.16}$$

where $i = 0, 1, 2, \ldots n$. While we postpone the case of third and higher order invariants to Sect. 3.4, the results given here are those for the first and second order invariants.

For the first order invariant, the PDE's to be solved for b_0 and b_1 are $(\partial b_1/\partial x) = 0$; $(\partial b_0/\partial x) = -(\partial b_1/\partial t)$; $\partial b_0/\partial t = b_1\,(\partial V/\partial x)$, which lead to the "potential" equation

$$(\partial V/\partial x) + (\ddot{\rho}_1/\rho_1)\,x - (\dot{\rho}_2/\rho_1) = 0, \tag{3.17}$$

with the only solution $V(x, t) = -(\ddot{\rho}_1/2\rho_1)\,x^2 + (\dot{\rho}_2/\rho_1)\,x + \rho_3(t)$. This system corresponds to a TD, rotating HO, expressed by $V(x, t) = (1/2)\,\omega^2(t)\,[x - \alpha(t)]^2$, and admits the invariant, $1 = \rho_2 + (\rho_1\dot{x} - \dot{\rho}_1\,x)$, where ρ_1 and ρ_2 are functions of t, satisfying

$$\ddot{\rho}_1 + \omega^2(t)\,\rho_1 = 0; \quad \dot{\rho}_2 + \omega^2(t)\,\alpha(t)\,\rho_1 = 0.$$

For the second order invariant, the PDE's to be solved for b_0, b_1 and b_2 are

$$(\partial b_2/\partial x) = 0\,;\, 2(\partial b_1/\partial x) + (\partial b_2/\partial t) = 0; \tag{3.16a, b}$$

$$(\partial b_0/\partial x) + (\partial b_1/\partial t) - b_2(\partial V/\partial x) = 0;\ (\partial b_0/\partial t) - b_1(\partial V/\partial x) = 0, \tag{3.16c, d}$$

and the "potential" equation turns out to be

$$\begin{aligned}&[-(1/2)\,\dot{\sigma}_1 x + \sigma_2]\,(\partial^2 V/\partial x^2) - \sigma_1(\partial^2 V/\partial t.\partial x)\\&- (3/2)\,\dot{\sigma}_1\,(\partial V/\partial x) + [-(1/2)\,\ddot{\sigma}_1 x + \ddot{\sigma}_2] = 0,\end{aligned} \tag{3.18}$$

where σ_1 and σ_2 are arbitrary functions of t and should be fixed by rationalizing (3.18) for a given V. Eq. (3.18) is a linear, second order PDE, whose solution, in principle, would provide the integrable systems admitting second order invariants. Using (3.18), while it is not difficult to recover the invariant (3.5) for the system (3.4), the case of a TD arbitrary power potential, namely $V(x, t) = \beta(t)\, x^m$, can also be analysed [73].

3.2.2 Ermakov Method

The study of a system of coupled, nonlinear second order oscillators possessing at least one invariant has become interesting from the point of view of applications. Ermakov [74] originally suggested a connection between the solutions of such a pair of coupled equations (hereafter termed as Eramakov systems). In recent years, Ray and Reid in a series of papers [68, 72, 75, 76] have studied these systems in the context of TDHO and with several degrees of generalizations. As a matter of fact, in their course of study, Ray and Reid have evolved a method of constructing the invariant for TD systems in 1D, known as Ermakov method and accordingly the invariant so constructed sometimes as the Ermakov invariant. This method, although simple but a heuristic one and sometimes leads to more general systems possessing invariants.

In this method, one eliminates $\omega^2(t)$ from the equation of motion for a TDHO, viz., (cf. system (3.4))

$$\ddot{x} + \omega^2(t)\, x = 0 \tag{3.19}$$

and the auxiliary eq. (3.6). As a result of the first integration of the resultant equation after multiplying the latter by $(\dot{x}\rho - x\dot{\rho})$, one immediately obtains the invariant (3.5). In this case, however, one has to know the auxiliary equation in advance. Besides accounting for the damping terms in (3.19) and (3.6), the most common generalization considered by Ray and Reid is in terms of the equations

$$\ddot{x} + \omega^2(t)\, x = g(\rho/x)/(x^2\rho), \tag{3.20a}$$

$$\ddot{\rho} + \omega^2(t)\, \rho = f\,(x/\rho)/(x\rho^2). \tag{3.20b}$$

These equations, as before, lead to the invariant

$$I = (1/2)\ [\phi(x/\rho) + \theta(\rho/x) + (x\dot{\rho}\text{-}\dot{x}\rho)^2], \tag{3.21}$$

with

$$\phi(x/\rho) = \int^{(x/\rho)} f(u)\, du\ :\ \theta(\rho/x) = 2 \int^{(\rho/x)} g(u)\, du.$$

We shall return to some of these discussions in Sect. 3.5.

3.2.3 Dynamical Algebraic Approach

Earlier Korsch [77] for a limited number of TD systems and later Kaushal and Korsch [66] for a variety of TD Hamiltonian systems exploited the closure property of dynamical Lie algebra generated by the phase-space functions, Γ's. Takayama [78] applied this approach to obtain the invariant for the system (3.10). While the details and extensions of this method to 2D systems will be discussed in the next Section, here we present the central idea.

In this approach one expresses the Hamiltonian of the system as

$$H = \sum_n h_n(t)\,\Gamma_n(x,p), \tag{3.22}$$

where the Γ_n's are not explicitly TD. Here the dynamical algebra is the Lie algebra of Γ_n's , which is closed with respect to the Poisson bracket,

$$[\Gamma_n\,,\Gamma_m]_{\mathrm{PB}} \;=\; \sum_r C^r_{nm}\;\Gamma_r\,, \tag{3.23}$$

where C^r_{nm} are the structure constants of the algebra. If the Γ_n's appearing in (3.22) are not sufficient to close the algebra then the set of Γ_n must be extended by the inclusion of new Γ_ℓ's such that $\Gamma_\ell = [\Gamma_n\ \Gamma_m]_{\mathrm{PB}}$ (with $h_\ell(t)$ taken to be zero in (3.22)), until the closure is obtained. It may be mentioned that the algebra contains the important structural information for the dynamical behaviour (independent of the particular functions $h_n(t)$ appearing in (3.22)) of the system besides its straightforward extension [64, 66] to the corresponding quantum case. Since the invariant, I, is also a phase-space function and so is the member of the dynamical algebra, it should be expressible as

$$I \;=\; \sum_k \lambda_k(t)\;\Gamma_k\,, \tag{3.24}$$

and its time-development should be in accordance with (1.14) and (1.15). Eqs. (1.14) and (1.15), in fact, imply

$$(\partial I/\partial t) = [H,\, I]_{\mathrm{PB}}, \tag{3.25}$$

which, after using (3.22) and (3.24), will lead [77] to the identity

$$\sum_r [\dot{\lambda}_r + \sum_{n,m} C^r_{nm}\, h_m(t)\,\lambda_n(t)\,]\,\Gamma_r \equiv 0\,,\ (r\;=\;1,2,\,\ldots).$$

Clearly, this identity provides a system of linear, first order ODE's, namely

$$\dot{\lambda}_r \;+\; \sum_n\;[\sum_m\; C^r_{nm}\;\; h_m(t)]\;\lambda_n \;=\; 0. \tag{3.26}$$

from which the unknown λ_k's can be determined. Once the λ_k's are known, the invariant can be computed from (3.24). For further details and applications of the method to specific examples we refer to Ref. [66].

3.2.4 Lutzky's Approach Using Noether's Theorem

This method, though analogous to the one discussed in Chapter 2 as Lie symmetries approach, is based on the following formulation of the Noether's theorem due to Lutzky [79] and subsequently used by Ray and Reid [72] for the TD systems in 1D. In this approach, modified for the TD case, the symmetry transformation is described by the group operator

$$X = \xi\,(x, t)\,\frac{\partial}{\partial t} + \eta\,(x, t)\,\frac{\partial}{\partial x}\,. \tag{3.27}$$

If the symmetry transformation defined by (3.27) leaves the action

$$A = \int L(x, \dot{x}, t)\; dt,$$

invariant, then the combination of the terms $\xi\,(\partial L/\partial t) + \eta(\partial L/\partial x) + (\dot{\eta} - \dot{x}\,\dot{\xi})\,(\partial L/\partial \dot{x}) + \dot{\xi}L$, is a total time derivative of a function $f(x, t)$, i.e.,

$$\xi\,(\partial L/\partial t) + \eta(\partial L/\partial x) + (\dot{\eta} - \dot{x}\,\dot{\xi})\,(\partial L/\partial \dot{x}) + \dot{\xi}\,L = \dot{f}\,. \tag{3.28}$$

It follows from this that a constant of motion for the system is

$$I = (\xi\,\dot{x} - \eta)\,(\partial L/\partial \dot{x}) - \xi L + f. \tag{3.29}$$

In (3.28) $\dot{\xi}$, $\dot{\eta}$ and $\dot{f}$ are defined as

$$\dot{\xi} = (\partial \xi\,/\,\partial t) + \dot{x}\,(\partial \xi\,/\,\partial x)\,;\; \dot{\eta} = (\partial \eta\,/\,\partial t) + \dot{x}\,(\partial \eta\,/\,\partial x),$$

$$\dot{f} = (\partial f/\partial t) + \dot{x}(\partial f/\partial x).$$

This method is successfully applied not only to TDHO [79] but also to several of its generalizations. The results derived are the same as obtained by using the dynamical algebraic approach.

3.2.5 Transformation–Group Method

With a view to obtaining an exact solution of the Schrodinger equation for a TDHO potential in 1D, transformation-group method was used by Ray [80]. This method, based on the transformation-group techniques introduced by Burgan *et. al.*, [81], essentially deals with the transformation of both dependent and independent variables. The unknown coefficient functions of the transformation are set in such a way that the form of the equation of motion remains invariant under the transformation. Interestingly, the energy-integral in the new coordinates turns out to be the desired invariant of the system. Here, we demonstrate the method for the system (3.10) and shall return in Chapter 5 for the details in the quantum context.

For the system (3.10), we use the transformation

$$x' = x / C(t) + A(t)\,;\, t' = D(t), \tag{3.30}$$

where C, A and D are the arbitrary function of t. Under this transformation, (3.10) takes the form [80]

$$C\dot{D}^2 \frac{d^2x'}{dt'^2} + (2\,\dot{C}\,\dot{D} + C\,\ddot{D})\frac{dx'}{dt'} + [\ddot{C} + \omega^2(t)\,C]\,x'$$

$$+ [-\ddot{C}\ A - 2\dot{C}\dot{A} - \omega^2(t)\ CA - C\ddot{A} - g] = 0. \tag{3.31}$$

Demanding that the form (3.10) remains invariant under (3.30), the coefficient of (dx' / dt') in (3.31) must vanish. This yields, $\dot{D} = dt' / dt = 1 / C^2$ and accordingly converts (3.31) into the form

$$\frac{d^2x'}{dt'^2} + C^3[\ddot{C} + \omega^2(t)\,C]x' + C^3\,[-\ddot{C}A - 2\dot{C}\dot{A} - \omega^2(t)\,CA - C\ddot{A} - g] = 0. \tag{3.32}$$

In order to identify (3.32) with the equation (i.e., with the equation of motion for a TID HO)

$$\frac{d^2\,x'}{dt'^2} + k\ x' = 0, \tag{3.33}$$

one has to chose A and C in (3.32) such that

$$\ddot{C} + \omega^2(t)\ C = k / C^3, \tag{3.34a}$$

$$\ddot{A} + (kA / C^4) + 2\,(\dot{C}\ \dot{A} / C) + g / C = 0. \tag{3.34b}$$

The energy integral for (3.33) has the form

$$I = (1/2)\ [(dx' / dt')^2 + k\,x'^2],$$

which, after carrying out the inverse transformation, reduces to the form

$$\mathrm{I} = (1/2)(C\dot{x} - \dot{C}\,x + C^2\,\dot{A})^2 + (1/2)\,k\,(x / C + A)^2, \tag{3.33$'$}$$

where C is any solution of (3.34a) and A is any solution of (3.34b). It may be mentioned that, the invariant (3.33′) and the eqs. (3.34) are the same as derived by Takayama [78] using the dynamical algebraic approach. Leach [69] has also employed the transformation (3.30) to find the invariants for some autonomous systems.

3.2.6 Other Methods

Besides the methods mentioned above, several other methods have also been used in the literature to construct the invariant for TD systems in 1D. In this regard, the method of self-similar techniques as used by Feix et al [82] is worth mentioning. The underlying idea of this method is rather simple and can be expressed as follows:

For the system (3.15) one looks for the invariant $I(x, p, t)$ in accordance with eqs. (1.14) and (1.15) which after using $\dot{x} = (\partial H / \partial p)$; $\dot{p} = -(\partial H / \partial x)$, reduce to the form

$$(\partial I / \partial t) + \dot{x}(\partial I / \partial x) + \dot{p}(\partial I / \partial p) = 0. \tag{3.35a}$$

Again, after using $\dot{x} = p$; $\dot{p} = -(\partial V / \partial x) = \mathcal{F}(x, t)$, this equation can be cast in the form

$$\mathcal{F}(x, t) = -[(\partial I / \partial t) + p\,(\partial I / \partial x)] / (\partial I / \partial p). \tag{3.35b}$$

Further, the differentiation of this equation w.r.t. p (keeping in mind that the left hand side is independent of p) leads to

$$(\partial I / \partial p)\,[(\partial^2 I / \partial t\,.\,\partial p) + (\partial I / \partial x) + p\,(\partial^2 I / \partial x\,.\,\partial p)]$$

$$-(\partial^2 I / \partial p^2)[(\partial I / \partial t) + p\,(\partial I / \partial x)] = 0. \tag{3.35c}$$

Now one assumes that the solution of (3.35c) are self-similar, i.e., one chooses to absorb the time in the reduced variables proportional to x and p in such a way that the (so called) self-similar transformation [82]

$$t = a^{\alpha}\,\bar{t}\ ;\ x = a^{\beta}\,\bar{x}\ ;\ p = a^{\gamma}\,\bar{p}\ ;\ I = a^{\delta}\,\bar{I}, \tag{3.35d}$$

leaves eq. (3.35c) unchanged and thus, yielding the relation $2\delta - \alpha - 2\gamma = 2\delta - \beta - \gamma$, and thereby leaving only three of α, β, γ and δ as arbitrary. Next, after introducing the parameters λ and μ through $\beta/\alpha = \lambda$, $\delta/\alpha = \mu$ and defining the new varaibles in terms of λ and μ as,

$$\xi = x / t^{\lambda}\ ;\ \eta = p / t^{\lambda-1}\ ;\ G = I / t^{\mu}, \tag{3.35e}$$

one can immediately write down the expressions for $\partial I / \partial t$, $\partial I / \partial x$ and $(\partial I / \partial p)$ in terms of these new variables. The use of these expressions in

(3.35b) gives rise to

$$\mathcal{F}(x, t) = -t^{\lambda-2}\,[\mu G + (\eta - \lambda\xi)\,(\partial G/\partial\xi) + (1 - \lambda)\eta.$$

$$(\partial G/\partial\eta)]\,/\,(\partial G/\partial\eta) = t^{\lambda-2}\,F(\xi), \tag{3.35f}$$

where $F(\xi)$ is the "reduced" force. One can define the "reduced" potential $\phi(\xi)$ through $F(\xi) = -\,(d\phi/d\xi)$, and express the potential term in (3.15) as

$$V(x, t) = t^{2\lambda-2}.\ \phi(x/t^{\lambda}). \tag{3.35g}$$

Finally, eq.(3.35f) can be cast in the form

$$\mu G + \omega(\partial G/\partial\xi) + [\tilde{F} + \omega(1 - 2\lambda)]\,(\partial G/\partial\omega) = 0, \tag{3.35h}$$

with $\omega = \eta - \lambda\xi$, and $\tilde{F} = F + \lambda(1 - \lambda)\,\xi$.

Since G is directly related to the invariant I (cf. eq., (3.35e)), the solution of (3.35h) will readily provide the invariant for the system (3.15). Not only this, different ansatze for the solution to eq.(3.35h), namely $G = \omega + a(\xi)$, $G = \omega^2 + a(\xi)\,\omega + b(\xi)$, $G = \omega^3 + a(\xi)\,\omega^2 + b(\xi)\,\omega + c(\xi), \ldots$, as considered by Feix et al [82] will give rise to invariants of different orders in momenta. Interestingly, the class of potentials $V(x, t) = U(x/C)/C^2 - (\ddot{C}/2C)\,x^2$, where $C(t)$ is a power function of t, is found to possess an energy-type invariant. More precisely, all potentials investigated using this method, finally, culminate to the form

$$V(x, t) = A\,(t^{2\lambda-2}(x/t^{\lambda})^{2\mu/K} + Bx^2/t^2, \tag{3.35i}$$

where A is arbitrary and K and B are again expressed in terms of λ and μ.

3.3 CONSTRUCTION OF THE FIRST INVARIANT IN TWO DIMENSIONS: VARIOUS METHODS

As mentioned before, the study of TD systems turns out to be more difficult than the TID systems in a given number of dimensions. For TD systems in 2D, while the available methods seem to be inadequate to provide the second invariant (in order to fulfil the integrabililty requirement) of the system, if it exists, not many attempts are made to obtain the invariants of order higher than the second order (in momenta). In this section, we present various methods used to study the first invariant of second order in 2D. In particular, the rationalization method and the dynamical algebraic approach will be discussed. Again, in the rationalization method both Cartesian and complex coordinate analysis will be carried out as was done for the TID systems in the previous chapter.

3.3.1 Rationalization Method

(a) Case of Cartesian Coordinates

The direct or rationalization method has been employed by several authors [83] to study the TD systems in 2D using the Cartesian coordinates. For the study of linear (in momenta) invariants we refer to these works. Here, in this section, we describe this method in the context of second order invariants only. For this purpose, we consider a dynamical system described by the Lagrangian,

$$L = (1/2)\ (\dot{x}_1^2\ +\ \dot{x}_2^2)\ -\ V(x_1,\ x_2,\ t), \tag{3.36}$$

and make an ansatz for the invariant, I, (cf. eq. (2.1)) as

$$I = a_o + a_i\xi_i + (1/2!)\ a_{ij}\ \xi_i\ \xi_j,\ (i,\ j = 1,2), \tag{3.37}$$

where $\xi_i = \dot{x}_i$; $a_{ij} = a_{ji}$ and a_o, a_i, a_{ij} are now functions of x_1, x_2 and t. Using eqs. (1.14) and (1.15) (with the Poisson bracket now defined for the 2D case) for the invariance of I and after accounting for the proper symmetrization, one arrives [20] at the following relations for the coefficient functions a_{ij}, a_i and a_o:

$$a_{ij,\,k} + a_{jk,\,i} + a_{ki,\,j} = 0, \tag{3.38}$$

$$a_{i,\,j} + a_{j,\,i} = -\,\partial a_{ij}\ /\ \partial t, \tag{3.39}$$

$$a_{o,\,i} + a_{ij}\ \dot{\xi}_j = -\,\partial a_i\ /\ \partial t, \tag{3.40}$$

$$a_j\ \dot{\xi}_j = -\partial a_o\ /\ \partial t. \tag{3.41}$$

In their detailed form these PDE's can be written as

$$(\partial a_{11}\ /\ \partial x_1) = 0;\ \ (\partial a_{22}\ /\ \partial x_2) = 0, \tag{3.42a, b}$$

$$2(\partial a_{12}\ /\ \partial x_1) + (\partial a_{11}\ /\ \partial x_2) = 0;\ 2(\partial a_{12}\ /\ \partial x_2) + (\partial a_{22}\ /\ \partial x_1) = 0, \tag{3.42c, d}$$

$$2(\partial a_1\ /\ \partial x_1) = -\,\partial a_{11}\ /\ \partial t\ ;\ \ 2(\partial a_2\ /\ \partial x_2) = -\ (\partial a_{22}\ /\ \partial t), \tag{3.42e, f}$$

$$(\partial a_1\ /\ \partial x_2)\ +\ (\partial a_2\ /\ \partial x_1)\ =\ -(\partial a_{12}\ /\ \partial t), \tag{3.42g}$$

$$\partial a_0\ /\ \partial x_1\ -\ a_{11}\ (\partial V\ /\ \partial x_1)\ -\ a_{12}(\partial V\ /\ \partial x_2) = -\ (\partial a_1\ /\ \partial t), \tag{3.42h}$$

$$\partial a_0\ /\ \partial x_2\ -\ a_{12}\ (\partial V\ /\ \partial x_1)\ -\ a_{22}(\partial V\ /\ \partial x_2) = -\ (\partial a_2\ /\ \partial t), \tag{3.42i}$$

$$a_1\ (\partial V\,/\,\partial x_1)\ +\ a_2\ (\partial V\,/\,\partial x_2)\ =\ (\partial a_0\ /\,\partial t). \tag{3.42j}$$

Note that the presence of the term linear in momenta in (3.37) leads to a larger number of equations here as compared to that of the TID case. Further, a simple analysis of eqs. (3.42a,b) and (3.42c,d) immediately leads [20] to the forms of a_{ij}'s as

$$a_{11}(x_2, t) = \psi_0(t)\, x_2^2 + \psi_2(t)\, x_2 + \psi_3(t),$$

$$a_{22}(x_1, t) = \psi_0(t)\, x_1^2 + \psi_1(t)\, x_1 + \psi_4(t), \tag{3.43}$$

$$a_{12}(x_1, x_2, t) = -\psi_0(t)\, x_1 x_2 - (1/2)[\psi_2(t)\, x_1 + \psi_1(t)\, x_2 - \mu(t)],$$

and subsequently the integration of eqs. (3.42e, f, g) yields for a_i's the expressions

$$\begin{aligned} a_1(x_1, x_2, t) &= -(1/2)[\dot{\psi}_2(t)\, x_2 + \dot{\psi}_3(t)]\, x_1 + (1/2)\dot{\psi}_1(t)\, x_2^2 \\ &\quad -(1/2)[\dot{\mu}(t) + \psi_5(t)]\, x_2 + (1/2)\, \psi_7(t), \\ a_2(x_1, x_2, t) &= -(1/2)[\dot{\psi}_1(t)\, x_1 + \dot{\psi}_4(t)]\, x_2 + (1/2)\dot{\psi}_2(t)\, x_1^2 \\ &\quad +(1/2)\, \psi_5(t)\, x_1 + (1/2)\, \psi_6(t). \end{aligned} \tag{3.44}$$

Here ψ_i's are the arbitrary functions of t. From now onwards we drop the arguments of the respective functions for the sake of brevity. Finally, after eliminating a_0 from eqs. (3.42h) and (3.42i) by differentiating them w.r.t. x_2 and x_1, respectively and subsequently using the results (3.43) and the fact that $(\partial^2 a_0 / \partial x_1\, \partial x_2) = (\partial^2 a_0 / \partial x_2\, \partial x_1)$, one arrives [20] at the following "potential" equation for the second order invariants.

$$\begin{aligned} &3[\ddot{\psi}_2\, x_2 - \ddot{\psi}_1\, x_1 + (1/2)\, \ddot{\mu} + \dot{\psi}_5] + 3(2\psi_0\, x_2 + \psi_2)\, (\partial V / \partial x_1) \\ &-3(2\psi_0 x_1 + \psi_1)\, (\partial V / \partial x_2) + (2\psi_0 x_1 x_2 + \psi_2 x_1 + \psi_1 x_2 - \mu). \\ &[(\partial^2 V / \partial x_1^2) - (\partial^2 V / \partial x_2^2)] + 2(\psi_0\, (x_2^2 - x_1^2) + \psi_2 x_2 - \psi_1 x_1 + \psi_3 - \psi_4). \\ &.(\partial^2 V / \partial x_1\, \partial x_2) = 0. \end{aligned} \tag{3.45}$$

This result was also derived by Gammaticos and Dorizzi [83] and used it to study a number of TD systems in 2D. It may be mentioned that the handling of this equation is even more difficult than the corresponding equation in the TID case (cf. eq. (2.2)). Nevertheless, we have made [20] an attempt to solve it in a general manner by considering separable forms of $V(x_1, x_2, t)$ in the variables x_1 and x_2, as before.

For the case when

$$V(x_1, x_2, t) = f(x_1, t) + g(x_2, t),$$

one of the special solutions of (3.45) yields the form of V as

$$V(x_1, x_2, t) = \psi_0[x_1 + \psi_1/2\psi_0]^2 + \psi_0[x_2 + \psi_2/2\psi_0]^2 + v_1(t), \tag{3.46}$$

which in analogy with the TID case, is the case of a *shifted rotating harmonic oscillator*. Here $v_1(t) = \psi_8 + \psi_9 - (\psi_1^2 + \psi_2^2)/2\psi_0$, is the pure TD part of V. Invariants are also constructed for several other special solutions of (3.45).

The rationalization of eq. (3.45) for a number of TD coupled oscillator systems has led to the construction of invariants for several cases. Here, however, we mention only the pertinent results.

(*i*) *TD Coupled oscillator:* For the system described by the potential

$$V(x_1, x_2, t) = \alpha_1(t)\, x_1^2 + \alpha_2(t)\, x_2^2 + \beta_1(t)\, x_1\, x_2, \qquad (3.47)$$

the invariant is obtained [20] as

$$I = [\alpha_1\, \psi_3 + (1/4)\, \ddot{\psi}_3]\, x_1^2 + [\alpha_2 \psi_3 + (1/4)\, \ddot{\psi}_3]\, x_2^2 + \beta_1 \psi_3\, x_1 x_2$$

$$+ (1/2)\, (\dot{\psi}_7\, x_1 - \dot{\psi}_6\, x_2) + (1/2)(\psi_7\, \dot{x}_1 + \psi_6\, \dot{x}_2) - (1/2)\, \dot{\psi}_3 (x_1 \dot{x}_1 + x_2 \dot{x}_2)$$

$$+ (1/2)\, c_5 (x_1 \dot{x}_2 - \dot{x}_1 x_2) + (1/2)\, \psi_3 (\dot{x}_1^2 + \dot{x}_2^2), \qquad (3.48)$$

where

$$\psi_3(t) = [c_5 / 2(\alpha_1 - \alpha_2)^{1/2}] \int (\beta_1 / (\alpha_1 - \alpha_2)^{1/2})\, dt,$$

with $c_5 \equiv$ const ; $\psi_6(t)$ and $\psi_7(t)$ are given by

$$\ddot{\psi}_6 = -2\, \psi_6\, \alpha_2 - \psi_7\, \beta_1\, ; \; \ddot{\psi}_7 = 2\, \psi_7\, \alpha_1 + \psi_6\, \beta_1,$$

and α_1, α_2 and β_1 satisfying a constraint

$$[\dot{\beta}_1 (\alpha_1 - \alpha_2) - \beta_1 (\dot{\alpha}_1 - \dot{\alpha}_2)]\, B_1 + 2\, (\alpha_1 - \alpha_2)^{1/2} [\beta_1^2 + (\alpha_1 - \alpha_2)^2] = 0, \qquad (3.49)$$

with $B_1 = \int (\beta_1 / (\alpha_1 - \alpha_2)^{1/2})\, dt$.

(*ii*) *TD Oscillator with inverse harmonic and cross terms:* In this case, we consider a general system described by the potential

$$V(x_1, x_2, t) = \alpha_1(t)\, x_1^2 + \alpha_2(t)\, x_2^2 + \beta(t)\, x_1^m\, x_2^n, \qquad (3.50)$$

and notice that the rationalization of the potential eq. (3.45) is possible only for the choice of the arbitrary functions as $\psi_0 = \psi_1 = \psi_2 = \mu = 0$, $\psi_3 = \psi_4$. Further, $\dot{\psi}_5 = 0$ implying $\psi_5 =$ const. (say c_5) and the numbers m and n must satisfy $m + n = -2$, as for the TID case (*cf.* Sect. 2.3.1). Though with this restriction on m and n several choices (including their fractional values) are possible, we give here results only for a few cases.

For the potential corresponding to $m = n = -1$, viz,

$$V(x_1, x_2, t) = \alpha_1(t)\, x_1^2 + \alpha_2(t)\, x_2^2 + \beta_0 x_1^{-1}\, x_2^{-1}, \qquad (3.51)$$

the invariant turns out to be

$$I = [\alpha_1\psi_3 + (1/4)\ddot{\psi}_3]\, x_1^2 + [\alpha_2\psi_3 + (1/4)\ddot{\psi}_3] x_2^2 + \beta_0\psi_3\, x_1^{-1}x_2^{-1}$$

$$-(1/2)\,\dot{\psi}_3(x_1\dot{x}_1 + x_2\dot{x}_2) + (1/2)\,\psi_3(\dot{x}_1^2 + \dot{x}_2^2), \qquad (3.52)$$

where ψ_3, and α_1 and α_2 satisfy the equations

$(1/4)\dddot{\psi}_3 + 2\alpha_1\,\dot{\psi}_3 + \dot{\alpha}_1\,\psi_3 = 0;\ (1/4)\dddot{\psi}_3 + 2\alpha_2\dot{\psi}_3 + \alpha_2\,\psi_3 = 0,$

and $c_5 = \psi_6 = \psi_7 = 0$. The above equations for ψ_3 yield $\psi_3(t) = c_3(\alpha_1 - \alpha_2)^{-1/2}$, and a constraint on α_1 and α_2 as

$$4(\ddot{\alpha}_1 - \ddot{\alpha}_2)(\alpha_1 - \alpha_2)^2 - 18(\alpha_1 - \alpha_2)(\dot{\alpha}_1 - \dot{\alpha}_2)(\ddot{\alpha}_1 - \ddot{\alpha}_2) + 15(\dot{\alpha}_1 - \dot{\alpha}_2)^3$$

$$= 32(\alpha_1 - \alpha_2)^3\,(\alpha_1\dot{\alpha}_2 - \dot{\alpha}_1\alpha_2). \qquad (3.53)$$

Further, a symmetrization of $x_1^m\ x_2^n$ – term in (3.50) with respect to m and n, namely the replacement of $\beta(t)\, x^m\, x^n$ by $(\beta_1(t)\ x_1^m\ x_2^n + \beta_2(t)\ x_1^n\ x_2^m)$ leads to several new results. For example, for $m = 0$, $n = -2$ and $m = -2$, $n = 0$ or for $m = 1$, $n = -3$ and $m = -3$, $n = 1$, the invariants corresponding to the TD oscillator with inverse harmonic or with cross terms can easily be derived [20]. Not only this, a criterion for the relative time-dependence of α_1, α_2 and β can be set. We shall return to some of these discussions in the context of dynamical algebraic approach.

(b) Case of Complex Coordinates

The complexification of two space dimensions in the form $Z = x_1 + ix_2$, $\bar{Z} = x_1 - ix_2$ for TD systems has also led to some interesting results as far as the construction of the first invariant is concerned. Particularly, the integrability of a variety of central potentials in this approach can be confirmed rather easily as compared to that in the Cartesian case. Here we extend the results of Sect. 2.3.1 to TD systems and restrict ourselves only to the construction of second order invariants in the ansatz (2.7). The systems we consider now are described by the Lagrangian,

$$L = (1/2)\,|\dot{Z}|^2 - V(Z,\bar{Z},t) \qquad (3.54)$$

with the corresponding equations of motion given by (2.6). We make [24] the same ansatz for I as (3.37), but now $\xi_1 = \dot{Z}$. $\xi_2 = \dot{\bar{Z}}$ and a_o, a_i, a_{ij} are the functions of Z, $\bar{Z}$ and t. Of course, the form of the equations satisfied by a_{ij}, a_j and a_o remain the same as (3.38) – (3.41), but their detailed version now turns [24] out to be as follows:

$$(\partial a_{11} / \partial Z) = 0; \; (\partial a_{22} / \partial \bar{Z}) = 0, \qquad (3.55a, b)$$

$$2(\partial a_{12} / \partial Z) + (\partial a_{11} / \partial \bar{Z}) = 0; \; 2(\partial a_{12} / \partial \bar{Z}) + (\partial a_{22} / \partial Z) = 0, \qquad (3.55c, d)$$

$$2(\partial a_1 / \partial Z) = -(\partial a_{11} / \partial t); 2(\partial a_2 / \partial \bar{Z}) = -(\partial a_{22} / \partial t), \qquad (3.55e, f)$$

$$(\partial a_1 / \partial \bar{Z}) + (\partial a_2 / \partial Z) = -(\partial a_{12} / \partial t), \qquad (3.55g)$$

$$(\partial a_0 / \partial Z) - 2a_{11} (\partial V / \partial \bar{Z}) - 2a_{12} (\partial V / \partial Z) = -(\partial a_1 / \partial t), \qquad (3.55h)$$

$$(\partial a_0 / \partial \bar{Z}) - 2a_{12} (\partial V / \partial \bar{Z}) - 2a_{22} (\partial V / \partial Z) = -(\partial a_2 / \partial t), \qquad (3.55i)$$

$$2a_1 (\partial V / \partial \bar{Z}) + 2a_2 (\partial V / \partial Z) = (\partial a_0 / \partial t). \qquad (3.55j)$$

As before, the solution of these PDE's will now lead [24, 25] to the following expressions for the coefficient functions a's,

$$a_{11} = c_1 \bar{Z}^2 + \psi_2 \bar{Z} + \psi_3; \; a_{22} = c_1 Z^2 + \psi_1 Z + \psi_4, \qquad (3.56a, b)$$

$$a_{12} = -c_1 Z\bar{Z} - (1/2)\psi_2 Z - (1/2)\bar{Z} + (1/2)\mu, \qquad (3.56c)$$

$$a_1 = -(1/2)[\dot{\psi}_2 \bar{Z} + \dot{\psi}_3] Z + (1/2)\dot{\psi}_1 \bar{Z}^2 - (1/2)[\dot{\mu} + \psi_5]\bar{Z} + \psi_7 / 2, \qquad (3.56d)$$

$$a_2 = -(1/2)[\dot{\psi}_1 Z + \dot{\psi}_4]\bar{Z} + (1/2)\dot{\psi}_2 Z^2 + (1/2)\psi_5 Z + \psi_6 / 2, \qquad (3.56e)$$

where ψ_i ($i = 1,....,7$) and μ are the arbitrary functions of t, and c_1 is, in general, a complex separation constant.

Now differentiating (3.55h) w.r.t. $\bar{Z}$ and (3.55i) w.r.t. Z and using $(\partial^2 a_0 / \partial Z \partial \bar{Z}) = (\partial^2 a_0 / \partial \bar{Z} \partial Z)$ and $(\partial^2 V / \partial Z \partial \bar{Z}) = (\partial^2 V / \partial \bar{Z} \partial Z)$ one arrives at

$$a_{22} (\partial^2 V / \partial Z^2) + [(\partial a_{22} / \partial Z) - (\partial a_{12} / \partial \bar{Z})] (\partial V / \partial Z) - a_{11} (\partial^2 V / \partial \bar{Z}^2)$$
$$+ [(\partial a_{12} / \partial Z) - (\partial a_{11} / \partial \bar{Z})] (\partial V / \partial \bar{Z}) + (1/2)[(\partial^2 a_1 / \partial t \partial \bar{Z})$$
$$-(\partial^2 a_2 / \partial t \, \partial Z)] = 0. \qquad (3.57)$$

Similarly, using $(\partial^2 a_0 / \partial Z.\partial t) = (\partial^2 a_0 / \partial t.\partial Z)$ and $(\partial^2 a_0 / \partial \bar{Z}.\partial t) = (\partial^2 a_o / \partial t.\partial \bar{Z})$ in eqs. (3.55h) and (3.55j), (3.55i) and (3.55j), respectively, one obtains another pair of equations, viz.,

$$a_2(\partial^2V/\partial Z^2)+[(\partial a_2/\partial Z)-(\partial a_{12}/\partial t)](\partial V/\partial Z)+a_1(\partial^2V/\partial\bar{Z}\partial Z)+$$

$$[(\partial a_1/\partial Z)-(\partial a_{11}/\partial t)](\partial V/\partial\bar{Z})-a_{11}(\partial^2V/\partial t\,\partial\bar{Z})-a_{12}(\partial^2V/\partial t\partial Z)+$$

$$(1/2)\ (\partial^2 a_1/\partial t^2)\ =\ 0, \tag{3.58}$$

$$a_1(\partial^2V/\partial\bar{Z}^2)+[(\partial a_1/\partial\bar{Z})-(\partial a_{12}/\partial t)](\partial V/\partial\bar{Z})+a_2(\partial^2V/\partial Z\,\partial\bar{Z})+$$

$$[(\partial a_2/\partial\bar{Z})-(\partial a_{22}/\partial t)](\partial V/\partial Z)-a_{22}(\partial^2V/\partial t\partial Z)-a_{12}(\partial^2V/\partial t\partial\bar{Z})+$$

$$(1/2)\ (\partial^2 a_2/\partial t^2)\ =\ 0, \tag{3.59}$$

In a way, eqs. (3.57) to (3.59) may constitute "potential" equations for the TD case. For a given form of V, while the rationalization of these three equations will fix some arbitrary functions involved in (3.56), the rest can be determined in the process of computing a_0 from (3.55j). Further note that for most of the systems eqs. (3.58) and (3.59) provide identical results.

For the choice $\ddot{\mu}(t)=0, \dot{\psi}_5(t)=0, \psi_1=\psi_2=\bar{\psi}_2, \psi_3=\psi_4=\bar{\psi}_4,$ eq. (3.57) yields [26, 27]

$$A(\partial^2V/\partial Z^2)\ +\ B(\partial V/\partial Z)\ +\ C\ =\ \bar{A}\ (\partial^2V/\partial\bar{Z}^2)+\bar{B}(\partial V/\partial\bar{Z})\ +\ \bar{C}$$

$$\equiv\ \phi(t)\ \text{(say)}, \tag{3.60}$$

where $A=2(c_1Z^2+\psi_1Z+\psi_3)$, $B=3(2c_1Z+\psi_1), C=-(3/2)\ddot{\psi}_2 Z.$

If $V(Z,\bar{Z},t)=V(|Z|,t)=\beta(t)\ v(|Z|)$, then the invariants are constructed [25] for the form $V(|Z|,t)=\beta(t)\ (b/r^4+d)$, (van der Waal-type potential), and $V(|Z|,t)=\beta(t)\ (\ln r+b_1/r^4+d_1)$ (confining-type potential). Other TD central potentials investigated are (i) the linearly confining potential

$$V(|Z|,\ t)=\omega(t)\ (Z.\bar{Z})^{1/2}-\beta(t)\ (Z.\bar{Z})^{-1/2}, \tag{3.61}$$

for which the invariant turns out to be [24]

$$I=\mu^{-1/2}\ (r-\mu/r)-(1/2)\ \dot{\mu}(x_1\dot{x}_1\ +\ x_2\dot{x}_2)\ -\ 2c_1(x_1\dot{x}_2-x_2\dot{x}_1)^2$$

$$+(1/2)\ \mu(\dot{x}_1^2+\dot{x}_2^2), \tag{3.62}$$

where $r^2\ =\ Z\bar{Z}$ and $\mu=a\,t^2+b't+c'$, and (ii) the harmonically confining potential

$$V(|Z|,t)\ \doteq\ -(1/2)\ (\ddot{u}/u)\ Z\bar{Z}\ -\ (\mu_0/u)\ \ (Z\bar{Z})^{-1/2}, \tag{3.63}$$

which is also studied by Katzin and Levine [84] using the method of symmetries. The invariant obtained for this latter potential is given by

$$I=[(\dot{\mu}/8\mu)r^2-(\sqrt{\mu}/\mu_0)r^{-1}]+k\mu_0(x_1x_2-x_2\dot{x}_1)[(\dot{\mu}/\sqrt{\mu})x_2-2\sqrt{\mu}\dot{x}_2]$$
$$-(1/\dot{2})\dot{\mu}(x_1\dot{x}_1+x_2\dot{x}_2)-2c_1\ (x_1\dot{x}_2-x_2\dot{x}_1)^2$$
$$+(1/2)\mu(\dot{x}_1^2+\dot{x}_2^2), \tag{3.64}$$

where $\mu = u^2/\mu_0^2$; $\psi_1 = \psi_2 = k\,u$, $\psi_5 = -(u\dot{u}/\mu_0^2)$. The invariant is also constructed for the TD Kepler potential, $V(|Z|, t) = -\beta(t)\,(Z\,\overline{Z})^{-1/2}$.

It is not difficult to extend to two dimensions the method of Lewis and Leach [83], which is based on an infinite series expansion of I in powers of the momentum. Further, correspondence of this method with the present one can be established [26] by writing the Poisson bracket of I and H as

$$[I,H]_{PB} = (\partial I/\partial Z)\,(\partial H/\partial \dot{Z}) - (\partial I/\partial \dot{Z})\,(\partial H/\partial Z) +$$
$$(\partial I/\partial \overline{Z})\,(\partial H/\partial \dot{\overline{Z}}) - (\partial I/\partial \dot{\overline{Z}})\,(\partial H/\partial \overline{Z}).$$

3.3.2 Dynamical Algebraic Approach

With a view to demonstrating the underlying elegance of the dynamical algebraic approach at the classical level, we present here its extended version to the 2D case which has been carried out recently [85]. No doubt the central idea of the method still remains the same, but now complexity of the algebra in terms of closure increases enormously.

The Hamiltonian for a 2D system, as before, can be expressed as

$$H = \sum_n h_n(t)\,\Gamma_n(x_1, p_1, x_2, p_z), \tag{3.65}$$

where the phase space functions, Γ_n, are the functions of x_1, p_1, x_2 and p_2 and they still close the Lie algebra through (3.23), but with respect to the Poisson bracket now defined as

$$[f, g]_{PB} = \frac{\partial f}{\partial x_1}\frac{\partial g}{\partial p_1} - \frac{\partial f}{\partial p_1}\frac{\partial g}{\partial x_1} + \frac{\partial f}{\partial x_2}\frac{\partial g}{\partial p_2} - \frac{\partial f}{\partial p_2}\frac{\partial g}{\partial x_2}. \tag{3.66}$$

The invariant I, also a member of the dynamical algebra, is now expressed as

$$I = \sum_k \lambda_k(t)\,\Gamma_k\,(x_1, p_1, x_2, p_2), \tag{3.67}$$

and fulfils the requirements (3.25), which finally leads to the set of equations similar to (3.26) for determining the unknown λ_k's. Here, we employ this method first to the simple case of coupled TD oscillators (3.47) and then to the case of its generalized version.

(*a*) *Coupled TD Oscillators:* The Hamiltonian corresponding to the potential (3.47) can be written as

$$H = (1/2)\,(p_1^2 + p_2^2) + \alpha_1(t)\, x_1^2 + \alpha_2(t)\, x_2^2 + \beta(t)\, x_1 x_2, \quad (3.68)$$

which we wish to express in the form (3.65) by identifying

$$\Gamma_1 = p_1^2 / 2,\ \Gamma_2 = p_2^2 / 2,\ \Gamma_3 = x_1^2,\ \Gamma_4 = x_2^2,\ \Gamma_5 = x_1 x_2, \quad (3.69a)$$

and

$$h_1 = h_2 = 1,\ h_3 = \alpha_1(t),\quad h_4 = \alpha_2(t),\ h_5 = \beta(t). \quad (3.69b)$$

In order to close the algebra, in this case, it becomes necessary to introduce five more Γ_k's. namely

$$\Gamma_6 = -2p_1x_1,\ \Gamma_7 = -p_1x_2,\ \Gamma_8 = -2p_2x_2,\ \Gamma_9 = -p_2x_1,\ \Gamma_{10} = p_1p_2, \quad (3.69c)$$

with the corresponding $h_k(t) = 0$ in (3.65). Further, the number of nonvanishing Poisson brackets turns [85] out to be 28 and their use in (3.25) yields [66] the following set of first-order ODE's in λ's:

$$\dot{\lambda}_1 = 4\lambda_6\,;\ \dot{\lambda}_2 = 4\lambda_8\,;\ \dot{\lambda}_3 = -4\alpha_1\lambda_6 - \beta\lambda_9\,;\ \dot{\lambda}_4 = -\beta\lambda_7 - 4\alpha_2\lambda_8, \quad (3.70a,\ b,\ c,\ d)$$

$$\dot{\lambda}_5 = -2\beta\lambda_6 - 2\alpha_1\lambda_7 - 2\beta\lambda_8 - 2\alpha_2\lambda_9\,;\ \dot{\lambda}_6 = -\alpha_1\lambda_1 + \lambda_3 - (1/2)\,\beta\lambda_{10}, \quad (3.70e,\ f)$$

$$\dot{\lambda}_7 = -\beta\lambda_1 + \lambda_5 - 2\alpha_2\lambda_{10}\,;\ \dot{\lambda}_8 = -\alpha_2\lambda_2 + \lambda_4 - (1/2)\,\beta\lambda_{10}, \quad (3.70g,\ h)$$

$$\dot{\lambda}_9 = -\beta\lambda_2 + \lambda_5 - 2\alpha_1\lambda_{10}\,;\ \dot{\lambda}_{10} = \lambda_7 + \lambda_9. \quad (3.70i,\ j)$$

As such the solution of these 10 coupled equations is difficult, but if we set λ_{10} const. (say k), $\lambda_1 = \lambda_2 = \psi(t)$, $\lambda_7 = -\lambda_4 = \eta(t)$ (say), then eqs. (3.70) can be solved immediately. As a result, the invariant (3.67) for the system (3.68) takes [85] the form

$$I = (1/2)\,\psi\,(p_1^2 + p_2^2) + [(1/4)\,\ddot{\psi} + \alpha_1\psi + (1/2)k\beta]\,x_1^2 +$$

$$[(1/4)\,\ddot{\psi} + \alpha_2\psi + (1/2)k\beta]\,x_2^2 + (\beta\,\psi + k\alpha_1 + k\alpha_2)x_1x_2$$

$$-(1/2)\dot{\psi}\,(x_1p_1 + x_2p_2) + \eta(x_1p_2 - x_2p_1) - k\,p_1p_2, \tag{3.71}$$

where ψ, η, α_1, α_2 and β satisfy the relations:

$$2\beta\dot{\psi} + \dot{\beta}\psi + k\,(\dot{\alpha}_1 + \dot{\alpha}_2) = -2(\alpha_1 - \alpha_2)\eta;\, \dot{\eta} = k(\alpha_1 - \alpha_2). \tag{3.72}$$

Note that when $k = 0$ (or η = const.), the result (3.71) reduces to (3.48) but only after setting $\psi_6 = \psi_7 = 0.$ in the latter.

(*b*) *Generalized TD oscillators:* Now, we consider a generalized form of (3.68) in which the coupling term $\beta(t)\,x_1x_2$ is replaced by an arbitrary function $\beta(t)\phi$, namely

$$H = (1/2)\,(p_1^2 + p_2^2) + \alpha_1(t)x_1^2 + \alpha_2(t)x_2^2 + \beta(t)\,\phi(x_1, x_2),$$

$$\equiv \Gamma_1 + \Gamma_2 + \alpha_1(t)\,\Gamma_3 + \alpha_2(t)\,\Gamma_4 + \beta(t)\,\Gamma_5, \tag{3.73}$$

with Γ_1, Γ_2, Γ_3, Γ_4 as defined before (cf. eq. (3.69a)) and $\Gamma_5 = \phi(x_1, x_2)$. As a result of this new definition of Γ_5, the affected (nonvanishing) Poisson brackets can be computed [85] as

$$[\Gamma_1, \Gamma_5]_{PB} = -p_1\,\frac{\partial\phi}{\partial x_1}\,;\; [\Gamma_2, \Gamma_5]_{PB} = -p_2\,\frac{\partial\phi}{\partial x_2},$$

$$[\Gamma_5, \Gamma_6]_{PB} = -2x_1\,\frac{\partial\phi}{\partial x_1}\,;\; [\Gamma_5, \Gamma_8]_{PB} = -2x_2\,\frac{\partial\phi}{\partial x_2}. \tag{3.74}$$

In this case, Γ_7 and Γ_9 are absent from the Lie algebra and Γ_{10} defined in the earlier case from $[\Gamma_2, \Gamma_7]_{PB} = \Gamma_{10}$, is also absent. Finally, one is left here only with 7 coupled equations in λ's, namely

$$\dot{\lambda}_1 = 4\lambda_6\,;\, \dot{\lambda}_2 = 4\lambda_8\,;\dot{\lambda}_3 = -4\,\alpha_1\lambda_6;\dot{\lambda}_4 = -4\,\alpha_2\lambda_8, \tag{3.75a, b, c, d}$$

$$\dot{\lambda}_5\phi = (\beta\lambda_1 - \lambda_5)p_1\,\frac{\partial\phi}{\partial x_1} + (\beta\lambda_2 - \lambda_5)\,p_2\,\frac{\partial\phi}{\partial x_2}$$

$$-\,2\beta\lambda_6 x_1\,\frac{\partial\phi}{\partial x_1} - 2\beta\lambda_8\,x_2\,\frac{\partial\phi}{\partial x_2}, \tag{3.75 e}$$

$$\dot{\lambda}_6 = -\,\alpha_1\lambda_1 + \lambda_3\,;\, \dot{\lambda}_8 = -\,\alpha_2\,\lambda_2 + \lambda_4. \tag{3.75f, g}$$

As before, here we again make an ansatz $\lambda_1 = \lambda_2 = \psi(t)$, and obtain the solution for other λ's from (3.75). This will give rise to the constraining relations

$$(\dddot{\psi} / 4) + 2\alpha_1\, \dot{\psi} + \dot{\alpha}_1\, \psi = 0, \; (\dddot{\psi} / 4) + 2\alpha_2\, \dot{\psi} + \dot{\alpha}_2\, \psi = 0, \qquad (3.76a, b)$$

and the form of the ϕ-equation (3.75e) as

$$\dot{\lambda}_5\phi - (\beta\psi - \lambda_5)\left(p_1 \frac{\partial\phi}{\partial x_1} + p_2 \frac{\partial\phi}{\partial x_2}\right) + (1/2)\, \beta\dot{\psi}\left(x_1 \frac{\partial\phi}{\partial x_1} + x_2 \frac{\partial\phi}{\partial x_2}\right) = 0. \qquad (3.77)$$

For the case when

$$\lambda_5 = \beta\psi, \qquad (3.78)$$

two particular solutions of (3.77) (namely, the ones separable in x_1 and x_2 coordinates under addition and multiplication operations) which lead [85] to the interesting cases are

$$\text{(i) } \phi(x_1, x_2) = k_1 x_1^{-\delta} + k_2 x_2^{-\delta}, \text{and (ii) } \phi(x_1, x_2) = k_3 (x_1 / x_2)^{c_1}\, x_1^{-\delta} + k_4 (x_2 / x_1)^{c_1} x_2^{-\delta}. \qquad (3.79a, b)$$

Here, c_1 and k_i (i = 1, 2, 3, 4) are the separation and integration constants, respectively and the function $\delta(t)$ is given by

$$\delta(t) = 2 + \dot{\beta}\psi / (\beta\, \dot{\psi}), \qquad (3.80a)$$

which, after using the form $\psi = c_0(\alpha_1 - \alpha_2)^{-1/2}$ (cf. eqs. (3.76)), reduces to

$$\delta(t) = 2 - \dot{\beta}(\alpha_1 - \alpha_2) / \beta(\dot{\alpha}_1 - \dot{\alpha}_2). \qquad (3.80b)$$

Now, it is not difficult to write down the invariants for the systems corresponding to the cases (3.79a) and (3.79b), which, respectively turn out to be

$$I = (1/2)\, \psi(p_1^2 + p_2^2) + [(1/4)\, \ddot{\psi} + \alpha_1\psi]\, x_1^2 + [(1/4)\, \ddot{\psi} + \alpha_2\psi] x_2^2 + \beta\psi\, [k_1\, x_1^{-\delta} + k_2\, x_2^{-\delta}] - (1/2)\, \dot{\psi}\, (x_1 p_1 + x_2 p_2),$$

and

$$I = (1/2)\, \psi(p_1^2 + p_2^2) + [(1/4)\, \ddot{\psi} + \alpha_1\psi]\, x_1^2 + [(1/4)\, \ddot{\psi} + \alpha_2\psi] x_2^2 + \beta\psi\, [k_3\, (x_1 / x_2)^{c_1} x_1^{-\delta} + k_4\, (x_2 / x_1)^{c_1} x_2^{-\delta}] - (1/2)\, \dot{\psi}\, (x_1 p_1 + x_2 p_2).$$

No doubt, the solution of the PDE (3.77) can suggest further examples of the systems admitting the first invariant, but what is of importance here is the rationale suggested by the present approach in terms of this equation regarding the relative time-dependence of the couplings in (3.73) vis-a-vis the time-dependence of the exponent δ. In this connection the following remarks are in order:

Eq. (3.80b) implies that the time-dependence in $\delta(t)$ arises mainly from the fact that $\alpha_1(t) \neq \alpha_2(t)$ in (3.73) i.e., for the oscillators with unequal spring constants. Alternatively, if $\dot{\beta}\psi = 0$ in (3.80a), then δ becomes independent of t and attains the value $\delta = 2$. As a result, since $\psi \neq 0$, $\dot{\beta}$ must be zero, thereby implying $\beta(t) = \text{const.}$ (say β_0). This will lead to the systems

$$V(x_1, x_2, t) = \alpha_1(t)\, x_1^2 + \alpha_2(t)\, x_2^2 + \beta_{10}\, x_1^{-2} + \beta_{20}\, x_2^{-2} \tag{3.81}$$

and

$$V(x_1, x_2, t) = \alpha_1(t)\, x_1^2 + \alpha_2(t)\, x_2^2 + \beta_{10}\, (x_1 / x_2)^{q_1}\, x_1^{-2} + \beta_{20}\, (x_2 / x_1)^{q_1}\, x_2^{-2}, \tag{3.82}$$

for which the invariants can also be constructed using the rationalization method. In fact, the system (3.82) is of special interest from the point of view of generalizing the Ermakov systems (cf. Sect. 3.5). By writing the cross term in (3.82) as $\phi = \beta_{10}\ x_1^m x_2^n + \beta_{20}\ x_1^n\ x_2^m$, one can notice that $m + n = -2$. This is another important rationale for the purposes of choosing the coupling terms in the systems admitting the quadratic invariants, and valid for both TD and TID (cf. Sect. 2.3) systems.

It is worth noting here that the coupled oscillator system studied above in case (a) (cf. eq. (3.68)) is not recoverable as such from the present general form (3.73). However, if one chooses the arbitrary function $\eta(t)$ and the constant k as zero in (3.71), then the structure of the invariant (3.71) reduces to either of the present forms of the invariant. Thus, it appears that the presence of the unknown function ϕ in (3.73) while gives more freedom towards the choice of the functional form of the coupling term, it however limits the analysis in terms of the arbitrary constants or TD functions.

(c) Generalized time- and momentum-dependent oscillators:

For a time- and momentum-dependent Hamiltonian system of the form

$$H = \frac{1}{2}(p_1^2 + p_2^2) + \alpha_1(t)\, x_1^2 + \alpha_2(t)\, x_2^2 + \beta(t)\, g(p_1, p_2),$$

$$\equiv \Gamma_1 + \Gamma_2 + \alpha_1(t)\ \Gamma_3 + \alpha_2(t)\, \Gamma_4 + \beta(t)\ \Gamma_5\,,$$

where Γ_1, Γ_2, Γ_3 and Γ_4, are the same as defined before and $\Gamma_5 = g(p_1, p_2)$, one needs two more Γ_i's, namely, $\Gamma_6 = -2p_1x_1$ and $\Gamma_7 = -2p_2x_2$ with $h_6 = h_7 = 0$, to close the algebra. Subsequently, one can easily compute the nonvanishing Poisson brackets. In this case while equations for other λ_i's remain the same, the equation for λ_5 (analogous to (3.75e)), however, turns out to be

$$\dot{\lambda}_5\, g(p_1, p_2) = -2\lambda_3\beta x_1\,(\partial g/\partial p_1) - 2\lambda_4\beta x_2\,(\partial g/\partial p_2) + 2\lambda_5\,[\alpha_1 x_1\,(\partial g/\partial p_1) + \alpha_2 x_2\,(\partial g/\partial p_2)] + 2\beta\,[\lambda_6\, p_1\,(\partial g/\partial p_1) + \lambda_7\, p_2\,(\partial g/\partial p_2)].$$

This equation for $\alpha_1 = \alpha_2 = \alpha$ (say) and $\lambda_5 = \beta\psi + (\beta\,\ddot{\psi}/4\alpha)$ reduces (see KPGM [85]) to the simpler form $\Delta(t)\, g(p_1, p_2) = p_1(\partial g/\partial p_1) + p_2(\partial g/\partial p_2)$, where $\Delta(t) = 2 + 2\,(\dot{\beta}\psi/\beta\dot{\psi}) + (\dddot{\psi}/2\alpha\dot{\psi}) + \ddot{\psi}\,(\dot{\beta}\alpha - \dot{\alpha}\beta)/(2\beta\alpha^2\dot{\psi})$. The ansatz for the solution of this PDE in the separable forms $g(p_1, p_2) = k_3\, p_1^{\Delta} + k_4\, p_2^{\Delta}$ and $g(p_1, p_2) = k_5\, p_1^{-c_1+\Delta}\, p_2^{c_1}$, respectively leads (see KPGM [85]) to the invariants

$$I = \frac{1}{2}\,\psi\,(p_1^2 + p_2^2) + \left(\frac{1}{4}\,\ddot{\psi} + \alpha\psi\right)(x_1^2 + x_2^2) + \left(\frac{\beta}{4\alpha}\,\ddot{\psi} + \beta\psi\right)(k_3 p_1^{\Delta} + k_4\, p_2^{\Delta}) - \frac{1}{2}\,\dot{\psi}\,(p_1x_1 + p_2x_2),$$

and

$$I = \frac{1}{2}\,\psi\,(p_1^2 + p_2^2) + \left(\frac{1}{4}\,\ddot{\psi} + \alpha\psi\right)(x_1^2 + x_2^2) + \left(\frac{\beta}{4\alpha}\,\ddot{\psi} + \beta\psi\right). (k_5\, p_1^{-c_1+\Delta}\, p_2^{c_1} + k_6\, p_1^{c_1}\, p_2^{-c_1+\Delta}) - \frac{1}{2}\,\dot{\psi}\,(p_1x_1 + p_2x_2),$$

as before. Here $\psi(t)$ satisfies the same equation as (3.76).

3.3.3 Painleve Method

There are not many applications of the Painleve conjecture (see, for example, Ramani et.al [51] to study the TD systems in 2D. However, the applications seem to be straightforward and can be carried out more or less in the same manner as it is done for the TID systems (cf. Sect. 2.3).

3.4 THIRD AND HIGHER ORDER INVARIANTS

In view of the fact that there exist additional difficulties in dealing with the systems involving explicit time-dependence, not many attempts have been made to construct the invariants of order higher than two. Here, we present a brief derivation of some results in the form of PDE's whose solutions would directly provide the systems admitting the third or fourth order invariants in both one and two dimensions. No doubt the method of Feix et al [82], based on the use of self-similar techniques, also provides the invariants of higher order in momenta, but it seems that only a restricted class of potentials (cf. Sect. 3.2.6) can be studied using this method. Therefore, we resort here to continue with the more general method, namely, the rationalization method.

3.4.1 Higher Order Invariants in One Dimension

We consider the system (3.15) and use the ansatz (3.14) along with the recursion relation (3.16).

(a) Third order invariants: For this case, in addition to eqs. (3.16c) and (3.16d), we also have [73] other PDE's from (3.16) as

$$(\partial b_3 / \partial x) = 0 \; ; 3(\partial b_2 / \partial x) + (\partial b_3 / \partial t) = 0, \qquad (3.83a, b)$$

$$2(\partial b_1 / \partial x) + (\partial b_2 / \partial t) - b_3(\partial V / \partial x) = 0. \qquad (3.83c)$$

The first two equations, respectively provide

$$b_3 = \psi_1(t), \; b_2 = -(1/3)\, \dot{\psi}_1\, x + \psi_2(t). \qquad (3.84)$$

Substituting these results for b_3 and b_2 in (3.83c) and after integrating the resultant equation one obtains

$$b_1 = (1/2)\, \ddot{\psi}_1\, x^2 - (1/2)\, \dot{\psi}_2 x + (1/2)\, \psi_1 V + \psi_3(t), \qquad (3.84')$$

where ψ_i's are some arbitrary functions of t. Using these results for b_1 and b_2, eqs. (3.16c) and (3.16d) can be used to eliminate b_0 in favour of V by noting the fact that $(\partial^2 b_0 / \partial x\, \partial t) = (\partial^2 b_0 / \partial t\, \partial x)$. This will lead to a general "potential" equation for the third order invariants

$$(1/2)\, \psi_1 V(\partial^2 V / \partial x^2) + [(1/12)\, \ddot{\psi}_1\, x^2 - (1/2)\, \dot{\psi}_2\, x + \psi_3]\, (\partial^2 V / \partial x^2)$$

$$+ [(1/2)\, \psi_1\, (\partial V / \partial x)^2 + (1/2)\, (\ddot{\psi}_1 x - 3\, \dot{\psi}_2)\, (\partial V / \partial x) +$$

$$[(1/3)\,\dot{\psi}_1 x - \psi_2].\,(\partial^2 V/\partial x.\partial t) + (1/2)\,\psi_1\,(\partial^2 V/\partial t^2) + \dot{\psi}_1\,(\partial V/\partial t) +$$

$$(1/2)\,\ddot{\psi}_1 V + [(1/12)\,\dddot{\psi}_1\,x^2 - (1/2)\,\ddot{\psi}_2\,x - \ddot{\psi}_3] = 0. \qquad (3.85)$$

Clearly the solution of this nonlinear equation is difficult. However, we notice from (3.84′) that $\psi_1(t) = 2(\partial^3 b_1/\partial x^3)/(\partial^3 V/\partial x^3)$, which implies the separability of b_1 and also of V in x and t variables to the extent that the ratio $(\partial^3 b_1/\partial x^3)/(\partial^3 V/\partial x^3)$ is only a function of t. For this reason we choose

$$b_1\,(x, t) = f(t)\,v(x),\; V(x, t) = (2f(t)/\psi_1)\,v(x). \qquad (3.86)$$

Such a choice would reduce [73] the potential eq. (3.85) to a simpler form

$$(1/v)[(d(vdv/dx/dx)) + [\dot{f}\,\psi_1 - c_3 f)\,((1/3)\,c_3 x - c_4)/\psi_1 f^2].$$

$$(1/v)(dv/dx) + (\psi_1\,\ddot{f}/2f^2) = 0, \qquad (3.87)$$

where c_3 and c_4 are arbitrary constants of integration. Finally, an integrable system of the form

$$V(x, t) = a\,t^{-4/3}\,(k_4 x + k_5)^{1/2}, \qquad (3.88)$$

admitting the invariant

$$I = -(1/2)\,[f - (k_4\,x + k_5)^{1/2}\,\dot{x}]^2 + (1/6)\,\psi_1\,\dot{x}^3 \qquad (3.89)$$

is obtained [73]. Here $f(t) \sim t^{1/3}$ and $\psi_1(t) \sim t$.

(b) *Fourth order invariants:* In this case, in addition to eqs. (3.16c), (3.16d) and (3.83c), from (3.16) we also have

$$(\partial b_4/\partial x) = 0\,; 4\,(\partial b_3/\partial x) + (\partial b_4/\partial t) = 0,$$

$$3\,(\partial b_2/\partial x) + (\partial b_3/\partial t) - b_4\,(\partial V/\partial x) = 0. \qquad (3.90\text{a, b, c})$$

As before, eqs. (3.90a) and (3.90b), lead to

$$b_4 = s_1(t) \text{ and } b_3 = -\,(1/4)\,\dot{s}_1\,x + s_2(t), \qquad (3.91\text{a, b})$$

which in turn provide the solution of (3.90c) as

$$b_2 = (1/24)\,\ddot{s}_1\,x^2 - (1/3)\,\dot{s}_2\,x + (1/3)\,s_1 V + s_3(t), \qquad (3.91\text{c})$$

where s_i's are some arbitrary functions of t. As before, using these expressions for b_2, b_3 and b_4 the integration of (3.83c) yields

$$b_1 = (-1/144)\,\dddot{s}_1\,x^3 + (1/12)\,\ddot{s}_2\,x^2 - (1/24)\,\dot{s}_1\gamma - (1/6)\,s_1(\partial\gamma/\partial t)$$
$$- (1/2)\,\dot{s}_3\,x - (1/2)\,[(1/4)\,\dot{s}_1\,x - s_2]\,V + s_4(t), \tag{3.91d}$$

where $\gamma = \int V\,dx$ and s_4 is another arbitrary function of t. As for the third order case, b_0 from eqs. (3.16c) and (3.16d) can be eliminated in favour of V again by noting the fact that $(\partial^2 b_0/\partial x.\partial t) = (\partial^2 b_0/\partial t.\partial x)$. This will, in fact, determine the "potential" equation for the fourth order invariants,

$$[(1/144)\,s_1^{(3)}\,x^3 - (1/12)\,s_2^{(2)}\,x^2 + (1/2)\,\dot{s}_3 x - s_4 + (1/24)\,\dot{s}_1\gamma$$
$$+ (1/6)\,s_1\,(\partial\gamma/\partial t) + (1/2)\,((1/4)\,\dot{s}_1 x - s_2)V\;\;(\partial^2 V/\partial x^2)$$
$$+ (1/2)\,((1/4)\,\dot{s}_1 x - s_2)\,(\partial V/\partial x)^2) + (1/2)\,[(1/8)\,s_1^{(3)}\,x^2 - s_2^{(2)}\,x$$
$$+ 3s_3 + \dot{s}_1 V + s_1\,(\partial V/\partial t)]\,(\partial V/\partial x) + [(1/24)\,s_1^{(2)}x^2 - (1/3)\,\dot{s}_2 x$$
$$+ s_3 + (1/3)\,s_1 V]\,(\partial^2 V/\partial x.\partial t) + [(1/144)\,s_1^{(5)}x^3 - (1/12)\,s_2^{(4)}\,x^2$$
$$+ (1/2)\,s_3^{(3)}\,x - s_4^{(2)} + (1/24)\,s_1^{(3)}\,\gamma + (1/4)\,s_1^{(2)}\,(\partial\gamma/\partial t) + (3/8)$$
$$\dot{s}_1\,(\partial^2\gamma/\partial t^2) + (1/6)\,s_1\,(\partial^3\gamma/\partial t^3) + (1/2)\,((1/4)\,s_1^{(3)}\,x - s_2^{(2)})\,V$$
$$+ ((1/4)\,s_1^{(2)}x - \dot{s}_2)\,(\partial V/\partial t) + (1/2)\,((1/4)\,\dot{s}_1 x - s_2)\,(\partial^2 V/\partial t^2)] = 0, \tag{3.92}$$

where the numbers in the parenthesis on the superscripts of s_i's represent the order of time-derivatives of s_i's. Note that the potential eq. (3.92) is a nonlinear, integro-PDE whose solution, in principle, would directly provide the integrable systems admitting the fourth order invariants.

As before, by writing $b_2(x, t)$ and $V(x, t)$ as separable functions in x and t variables as

$$b_2(x, t) = g(t)\,w(x),\; V(x, t) = (3g(t)/s_1)\,w(x), \tag{3.93}$$

the potential equation (3.92) can be expressed in a reduced form

$$(\dot{g}\,s_1 - (3/4)\,g\,\dot{s}_1)\,W\,w'' + 3g\,((1/4)\,\ddot{s}_1\,x - s_2)\,(w\,w')' +$$
$$(5\dot{g}\,s_1 - 2g\,\dot{s}_1)\,w\,w' + (1/g)((1/3)\,\dddot{g}\,s_1{}^2 - (1/4)\,\ddot{g}\,\dot{s}_1\,s_2 +$$
$$(1/2)\,\dot{g}\,\dot{s}_1^2 - \dot{s}_1^3\,g/2s_1)\,W + (1/g)\,((1/4)\,\dot{s}_1 x - s_2).$$
$$(\ddot{g}\,s_1 - 2\dot{s}_1\dot{g} + 2g\,\dot{s}_1^2/s_1)\,w = 0, \tag{3.94}$$

where $W = \int w\,dx$ and $s_4(t)$ is taken to be zero, and the primes on w

represent the derivatives of w w.r.t. x. Eq. (3.94) appears to be difficult to solve even for a trivial case like the one discussed in the third order case.

3.4.2 Third Order Invariants in Two Dimensions

As the complexity in the derivation of an invariant for TD systems increases with respect to both its order and the dimensionality of the system, here we present the derivation only of third order invariants and that too by using the rationalization method in terms of the complex coordinates. The corresponding results for the Cartesian case can be derived in an analogous manner.

We consider the system described by the Lagrangian (3.54) and now make an ansatz for the invariant as

$$I = a_0 + a_i\, \xi_i + (1/2)\, a_{ij}\, \xi_i\, \xi_j + (1/6)\, a_{ijk}\, \xi_i\, \xi_j\, \xi_k, \qquad (i, j, k = 1, 2), \qquad (3.95)$$

where $\xi_1 = \dot{Z}$ and $\xi_2 = \dot{\overline{Z}}$, and a_0, a_i, a_{ij}, a_{ijk} are the functions of Z, $\overline{Z}$ and t as before. For the invariance of I, one uses eqs. (1.14) and (1.15) and obtains an identity. Now equating the coefficients of the powers of ξ_1 and ξ_2 and their products to zero after accounting for the proper symmetrization in the resultant expression, one obtains the following relations for a's as before:

$$a_{ijk,\,l} + a_{jkl,\,i} + a_{kli,\,j} + a_{lij,\,k} = 0, \qquad (3.96)$$

$$a_{ij,k} + a_{jk,\,i} + a_{ki,\,j} + (\partial a_{ijk} / \partial t) = 0, \qquad (3.97)$$

$$a_{i,j} + a_{j,\,i} + (\partial a_{ij} / \partial t) + a_{ijk}\dot{\xi}_k = 0, \qquad (3.98)$$

$$a_{0,i} + \partial a_i / \partial t + a_{ij}\, \dot{\xi}_j = 0, \qquad (3.99)$$

$$(\partial a_0 / \partial t) + a_i\, \dot{\xi}_i = 0. \qquad (3.100)$$

Now, using $\dot{\xi}_1 = -2(\partial V / \partial \overline{Z}), \dot{\xi}_2 = -2(\partial V / \partial Z)$, eqs. (3.96) to (3.100) yield the following set of coupled PDE's:

$$(\partial a_{111} / \partial Z) = 0\,; \quad (\partial a_{222} / \partial \overline{Z}) = 0, \qquad (3.101\text{a, b})$$

$$(\partial a_{111} / \partial \overline{Z}) + 3(\partial a_{112} / \partial Z) = 0; \quad (\partial a_{112} / \partial \overline{Z}) + (\partial a_{122} / \partial Z) = 0, \qquad (3.101\text{c, d})$$

$$(\partial a_{222} / \partial Z) + 3(\partial a_{122} / \partial \overline{Z}) = 0; \quad 3(\partial a_{11} / \partial Z) + (\partial a_{111} / \partial t) = 0, \qquad (3.101\text{e, f})$$

$$(\partial a_{11}/\partial\bar{Z}) + 2(\partial a_{12}/\partial Z) + (\partial a_{112}/\partial t) = 0, \tag{3.101g}$$

$$(\partial a_{22}/\partial Z) + 2(\partial a_{12}/\partial\bar{Z}) + (\partial a_{122}/\partial t) = 0, \tag{3.101h}$$

$$3(\partial a_{22}/\partial\bar{Z}) + (\partial a_{222}/\partial t) = 0, \tag{3.101i}$$

$$2(\partial a_{1}/\partial Z) + (\partial a_{11}/\partial t) = 2a_{111}(\partial V/\partial\bar{Z}) + 2a_{112}(\partial V/\partial Z), \tag{3.101j}$$

$$2(\partial a_{2}/\partial\bar{Z}) + (\partial a_{22}/\partial t) = 2a_{122}(\partial V/\partial\bar{Z}) + 2a_{222}(\partial V/\partial Z), \tag{3.101k}$$

$$(\partial a_{1}/\partial\bar{Z}) + (\partial a_{2}/\partial Z) + \partial a_{12}/\partial t) = 2a_{112}(\partial V/\partial\bar{Z}) + 2a_{122}(\partial V/\partial Z), \tag{3.101l}$$

$$(\partial a_{0}/\partial Z) + (\partial a_{1}/\partial t) = 2a_{11}(\partial V/\partial\bar{Z}) + 2a_{12}(\partial V/\partial Z), \tag{3.101m}$$

$$(\partial a_{0}/\partial\bar{Z}) + (\partial a_{2}/\partial t) = 2a_{12}(\partial V/\partial\bar{Z}) + 2a_{22}(\partial V/\partial Z), \tag{3.101n}$$

$$(\partial a_{0}/\partial t) = 2a_{1}(\partial V/\partial\bar{Z}) + 2a_{2}(\partial V/\partial Z). \tag{3.101o}$$

Note that, for $a_{ijk}= 0$ in (3.95), these 15 equations reduce to 10 equations which are analysed earlier (cf. Sect. 3.3.1) for the second order case. Now, we present the solutions of these equations for determining the coefficient functions, a's.

(i) *Determination of* a_{ijk}: Clearly, eqs. (3.101a) and (3.101b) imply that $a_{111} \equiv a_{111}(\bar{Z},t) = \Psi_1(\bar{Z},t); a_{222} = a_{222}(Z,t) = \Phi_1(Z,t)$. On differentiating (3.101c) w.r.t. $\bar{Z}$ and using (3.101d), one obtains

$$(\partial^2 a_{111}/\partial\bar{Z}^2) - 3(\partial^2 a_{122}/\partial Z^2) = 0,$$

and similarly, the differentiation of (3.101e) w.r.t Z, yields

$$(\partial^2 a_{222}/\partial Z^2) + 3(\partial^2 a_{122}/\partial Z.\partial\bar{Z}) = 0.$$

Further differentiation of these two equations w.r.t. $\bar{Z}$ and Z, respectively, after noting the fact that $(\partial^3 a_{122}/\partial\bar{Z}.\partial Z^2) = (\partial^3 a_{122}/\partial Z^2.\partial\bar{Z})$, leads to

$$(\partial^3 a_{111}/\partial\bar{Z}^3) = -(\partial^3 a_{222}/\partial Z^3) = \text{a function of } t \text{ alone (say } \sigma_1(t)). \tag{3.102}$$

If we assume the separability of Ψ_1 and Φ_1 as $\Psi_1(\bar{Z}, t) = \psi_1(\bar{Z}) f_1(t)$; $\Phi_1(Z, t) = \phi_1(Z)\, g_1(t)$, then the solution to eqs. (3.102) can be obtained immediately as

$$a_{111} = (1/6)\,\sigma_1\,\bar{Z}^3 + (1/2)\,\sigma_2\,\bar{Z}^2 + \sigma_3\,\bar{Z} + \sigma_4; \tag{3.103a}$$

$$a_{222} = -(1/6)\,\sigma_1\,Z^3 + (1/2)\,\sigma_5\,Z^2 + \sigma_6\,Z + \sigma_7. \qquad (3.103b)$$

The coefficient functions a_{112} and a_{122} can be obtained in the same way from the integrations of (3.101c) and (3.101e) as

$$a_{112} = -(1/6)\,\sigma_1\bar{Z}^2 Z - (1/3)\,\sigma_2\bar{Z}Z + (1/6)\,\sigma_5\bar{Z}^2$$
$$-(1/3)\,\sigma_3 Z + \sigma_8\bar{Z} + \sigma_9, \qquad (3.103c)$$

$$a_{122} = (1/6)\,\sigma_1 Z^2\bar{Z} - (1/3)\,\sigma_5\bar{Z}Z + (1/6)\,\sigma_2 Z^2$$
$$-(1/3)\,\sigma_6\bar{Z} - \sigma_8 Z + \sigma_{10}. \qquad (3.103d)$$

Regarding the notations, note that throughout this subsection the arbitrary functions Ψ_i's are the functions of $\bar{Z}$ and t. and Φ_i's are those of Z and t; ψ_i's are the functions of $\bar{Z}$ alone and ϕ_i's are that of Z alone, and σ_i's are the functions of t alone. Further σ_1 is the separation function and other σ_i's are the arbitrary functions of integration.

(ii) Determination of a_{ij}: To obtain a_{ij} we use (3.103a) and (3.103b) in (3.101f) and (3.101i), respectively and by integrating the resultant equations one obtains the expressions for a_{11} and a_{22} as,

$$a_{11} = -(1/18)\,\dot{\sigma}_1 Z\bar{Z}^3 - (1/6)\,\dot{\sigma}_2 Z\bar{Z}^2 - (1/3)\,\dot{\sigma}_3 Z\bar{Z}$$
$$-(1/3)\,\dot{\sigma}_4 Z + \Psi_3(Z,t), \qquad (3.104a)$$

$$a_{22} = (1/18)\,\dot{\sigma}_1 Z^3\bar{Z} - (1/6)\,\dot{\sigma}_5 Z^2\bar{Z} - (1/3)\,\dot{\sigma}_6 Z\bar{Z}$$
$$-(1/3)\,\dot{\sigma}_7\bar{Z} + \Phi_3(Z,t). \qquad (3.104b)$$

These expressions when used respectively in (3.101g) and (3.101h) along with (3.103c) and (3.103d), will yield the two expressions for a_{12} involving some arbitrary functions Ψ_3, Ψ_4 and Φ_3, Φ_4. A comparison of these two expressions for a_{12} provides the identity

$$(1/6)\,\dot{\sigma}_1 Z^2\bar{Z}^2 - (1/4)\,\dot{\sigma}_5\bar{Z}^2 Z + (1/4)\,\dot{\sigma}_2 Z^2\bar{Z} + (1/6)\,\dot{\sigma}_3 Z^2 - (1/6)\,\dot{\sigma}_6\bar{Z}^2$$
$$-\dot{\sigma}_8 Z\bar{Z} - (1/2)\,\dot{\sigma}_9 Z + (1/2)\,\dot{\sigma}_{10}\bar{Z} - (1/2)\,Z\,(\partial\Psi_3/\partial\bar{Z}) + \Psi_4 + (1/2)\bar{Z}$$
$$(\partial\Phi_3/\partial Z) + \Phi_4 = 0. \qquad (3.105)$$

From this identity one immediately obtains

$$\dot{\sigma}_1 = 0 \text{ or } \sigma_1 = \text{const. (say } c_1) \qquad (3.106a)$$

and subsequently

$$(\partial^3 \Psi_3 / \partial \bar{Z}^3) = -(1/2)\,\dot{\sigma}_5 \;;\; (\partial^3 \Phi_3 / \partial Z^3) = -(1/2)\,\dot{\sigma}_2,$$

which imply

$$\Psi_3(\bar{Z}, t) = (-1/12)\,\dot{\sigma}_5 \bar{Z}^3 + (1/2)\,\sigma_{11}\bar{Z}^2 + \sigma_{12}\bar{Z} + \sigma_{13},$$

$$\Phi_3(Z, t) = (-1/12)\,\dot{\sigma}_2 Z^3 + (1/2)\,\sigma_{14} Z^2 + \sigma_{15} Z + \sigma_{16},$$

where from Ψ_4 and Φ_4 in (3.105) can be set as

$$\Psi_4(\bar{Z}, t) = (1/6)\,\dot{\sigma}_6 \bar{Z}^2 - (1/2)\,\dot{\sigma}_{10}\,\bar{Z};$$

$$\Phi_4(Z, t) = (1/6)\,\dot{\sigma}_3 Z^2 - (1/2)\,\dot{\sigma}_9\,Z.$$

With these results, the two expressions for a_{12} become

$$a_{12} = (1/6)\,\dot{\sigma}_2 \bar{Z} Z^2 + (1/24)\,\dot{\sigma}_5\,\bar{Z}^2 Z + (1/6)\,\dot{\sigma}_3 Z^2 + (1/6)\,\dot{\sigma}_6 \bar{Z}^2$$
$$-(1/2)\,(\dot{\sigma}_8 + \sigma_{11})\,Z\bar{Z} - (1/2)\,(\dot{\sigma}_9 + \sigma_{12})\,Z - (1/2)\,\dot{\sigma}_{10}\,\bar{Z}.$$

and

$$a_{12} = (1/24)\,\dot{\sigma}_2 \bar{Z} Z^2 + (1/6)\,\dot{\sigma}_5\,Z\bar{Z}^2 + (1/6)\,\dot{\sigma}_3 Z^2 + (1/6)\,\dot{\sigma}_6 \bar{Z}^2$$
$$+(1/2)\,(\dot{\sigma}_8 - \sigma_{14})\,Z\bar{Z} - (1/2)\,(\dot{\sigma}_{10} + \sigma_{15})\,\bar{Z} - (1/2)\,\dot{\sigma}_9 Z.$$

Uniqueness of these two expressions will further require that

$$\dot{\sigma}_2 = \dot{\sigma}_5 = 0\,;\, \sigma_{12} = \sigma_{15} = 0,\, 2\dot{\sigma}_8 + \sigma_{11} - \sigma_{14} = 0. \qquad (3.106\text{b, c, d})$$

Eqs. (3.106b) imply σ_2 = const. (say c_2) and σ_5 = const. (say c_5). Finally, the coefficients a_{11}, a_{22} and a_{12} become

$$a_{11} = -(1/3)\,\dot{\sigma}_3 Z\bar{Z} - (1/3)\,\dot{\sigma}_4 Z + (1/2)\,\sigma_{11}\bar{Z}^2 + \sigma_{13}, \qquad (3.107\text{a})$$

$$a_{22} = -(1/3)\,\dot{\sigma}_6 Z\bar{Z} - (1/3)\,\dot{\sigma}_7 \bar{Z} + (1/2)\,\sigma_{14} Z^2 + \sigma_{16}, \qquad (3.107\text{b})$$

$$a_{12} = (1/6)\,\dot{\sigma}_3 Z^2 + (1/6)\,\dot{\sigma}_6 \bar{Z}^2 - (1/2)\,\dot{\sigma}_9 Z - (1/2)\,\dot{\sigma}_{10}\,\bar{Z}, \qquad (3.107\text{c})$$

and the expressions for a_{ijk} from (3.103) now take the form

$$a_{111} = (1/6)\,c_1 \bar{Z}^3 + (1/2)\,c_2 \bar{Z}^2 + \sigma_3\,\bar{Z} + \sigma_4. \qquad (3.108\text{a})$$

$$a_{222} = -(1/6)\,c_1 Z^3 + (1/2)\,c_5 Z^2 + \sigma_6 Z + \sigma_7. \qquad (3.108\text{b})$$

$$a_{112} = -(1/6)c_1 \bar{Z}^2 Z - (1/3)\,c_2 \bar{Z} Z + (1/6)c_5 \bar{Z}^2 - (1/3)\sigma_3 Z + \sigma_8 \bar{Z} + \sigma_9,$$
$$(3.108\text{c})$$

$$a_{122} = (1/6)\, c_1 Z^2\bar{Z} - (1/3)c_5 Z\bar{Z} + (1/6)\, c_2 Z^2 - (1/3)\sigma_6\bar{Z} - \sigma_8 Z + \sigma_{10}. \tag{3.108d}$$

(iii) Derivation of the "potential" equations: For the terms involving the potential in eqs. (3.101), we introduce the following notations for convenience:

$$F = a_{112}(\partial V/\partial\bar{Z}) + a_{122}(\partial V/\partial Z)$$

$$G_1 = a_{122}(\partial V/\partial\bar{Z}) + a_{222}(\partial V/\partial Z);\, G_2 = a_{111}(\partial V/\partial\bar{Z}) + a_{112}(\partial V/\partial Z)$$

$$R = a_{11}(\partial V/\partial\bar{Z}) + a_{12}(\partial V/\partial Z);\; S = a_{12}(\partial V/\partial\bar{Z}) + a_{22}(\partial V/\partial Z). \tag{3.109}$$

Differentiating (3.101*l*) w.r.t $\bar{Z}$ and then using (3.101k) for $(\partial a_2/\partial\bar{Z})$ one obtains,

$$(\partial^2 a_1/\partial\bar{Z}^2) = -\partial[(\partial a_{12}/\partial\bar{Z}) - (1/2)\,(\partial a_{22}/\partial Z)]/\partial t + 2(\partial F/\partial\bar{Z})$$
$$-(\partial G_1/\partial Z),$$

which, after using (3.107b) and (3.107c) reduces to

$$(\partial^2 a_1/\partial\bar{Z}^2) = -(1/2)(\ddot{\sigma}_6\,\bar{Z} - \dot{\sigma}_{14}Z - \ddot{\sigma}_{10}) + 2(\partial F/\partial Z) - (\partial G_1/\partial Z). \tag{3.110}$$

Similarly eq. (3.101*l*), after using (3.101j), (3.107a) and (3.107c), becomes

$$(\partial^2 a_2/\partial Z^2) = -(1/2)\,(\ddot{\sigma}_3\, Z - \dot{\sigma}_{11}\bar{Z} - \ddot{\sigma}_9) + 2(\partial F/\partial Z) - (\partial G_2/\partial\bar{Z}). \tag{3.111}$$

Thus, the coefficients a_1 and a_2 can be computed by integrating (3.110) and (3.111), respectively.

On differentiating (3.101m) and (3.101n) w.r.t. $\bar{Z}$ and Z, respectively and then using $(\partial^2 a_0/\partial Z\,.\partial\bar{Z}) = (\partial^2 a_0/\partial\bar{Z}.\partial Z)$ for eliminating a_0, one obtains

$$\frac{\partial}{\partial t}(\partial a_1/\partial\bar{Z}) - \frac{\partial}{\partial t}(\partial a_2/\partial Z) = 2[(\partial R/\partial\bar{Z}) - (\partial S/\partial Z)]. \tag{3.112a}$$

Similarly, the differentiation of (3.101*l*) w.r.t. t gives

$$\frac{\partial}{\partial t}(\partial a_1/\partial\bar{Z}) + \frac{\partial}{\partial t}(\partial a_2/\partial Z) = -(\partial^2 a_{12}/\partial t^2) + 2(\partial F/\partial t). \tag{3.112b}$$

Now, eqs. (3.112a) and (3.112b) give rise to

$$\frac{\partial}{\partial t}(\partial a_1 / \partial \overline{Z}) = -(1/2)(\partial^2 a_{12} / \partial t^2) + (\partial F / \partial t) + [(\partial R / \partial \overline{Z}) - (\partial S / \partial Z)], \tag{3.113a}$$

$$\frac{\partial}{\partial t}(\partial a_2 / \partial Z) = -(1/2)(\partial^2 a_{12} / \partial t^2) + (\partial F / \partial t) - [(\partial R / \partial \overline{Z}) - (\partial S / \partial Z)]. \tag{3.113b}$$

In order to eliminate a_1 and a_2 from (3.113a) and (3.113b), we differentiate them w.r.t. Z and $\overline{Z}$, respectively and then correspondingly use the results (3.110) and (3.111). This will give [86] the following pair of "potential" equations:

$$(\partial^2 F / \partial t.\partial \overline{Z}) - (\partial^2 G_1 / \partial t.\partial Z) - (1/2)(\ddot{\sigma}_6 \overline{Z} - \ddot{\sigma}_{14} Z - \ddot{\sigma}_{10})$$
$$= (\partial / \partial \overline{Z})[(\partial R / \partial \overline{Z}) - (\partial S / \partial Z)] - (1/2)(\partial^3 a_{12} / \partial \overline{Z}.\partial t^2), \tag{3.114}$$

and

$$(\partial^2 F / \partial t.\partial Z) - (\partial^2 G_2 / \partial t.\partial \overline{Z}) - (1/2)(\ddot{\sigma}_3 Z - \ddot{\sigma}_{11} \overline{Z} - \ddot{\sigma}_9)$$
$$= -(\partial / \partial Z)[(\partial R / \partial \overline{Z}) - (\partial S / \partial Z)] - (1/2)(\partial^3 a_{12} / \partial Z.\partial t^2). \tag{3.115}$$

An alternative and somewhat simple looking forms of these equations can be obtained by differentiating (3.114) and (3.115) once again w.r.t. Z and $\overline{Z}$, respectively and then noting the facts that

$$\frac{\partial^2}{\partial Z.\partial \overline{Z}}[(\partial R / \partial \overline{Z}) - (\partial S / \partial Z)] = \frac{\partial^2}{\partial \overline{Z}.\partial Z}[(\partial R / \partial \overline{Z}) - (\partial S / \partial Z)],$$

and

$$(\partial^4 a_{12} / \partial Z.\partial \overline{Z}.\partial t^2) = (\partial^4 a_{12} / \partial \overline{Z}.\partial Z.\partial t^2) = 0, \quad \text{(cf. eq. (3.107c))}.$$

Thus, eqs. (3.114) and (3.115) reduce to

$$\frac{\partial}{\partial t}[(\partial^2 G_1 / \partial Z^2) + (\partial^2 G_2 / \partial \overline{Z}^2)] - 2(\partial^3 F / \partial t.\partial Z.\partial \overline{Z}) = (1/2)(\ddot{\sigma}_{11} + \ddot{\sigma}_{14}), \tag{3.114$'$}$$

$$\frac{\partial}{\partial t}[(\partial^2 G_1 / \partial Z^2) - (\partial^2 G_2 / \partial \overline{Z}^2)] + 2\frac{\partial^2}{\partial Z.\partial Z}[(\partial R / \partial \overline{Z}) - (\partial S / \partial Z)]$$
$$= -(1/2)(\ddot{\sigma}_{11} + \ddot{\sigma}_{14}). \tag{3.115$'$}$$

Recall that F, G_1, G_2, R and S involve derivatives of the potential function (cf. eqs. (3.109)) V. As a result, the solutions of eqs. (3.114) and (3.115) (or of eqs. (3.114') and (3.115')) would directly provide the systems admitting the third order first invariant. Further, these solutions have to be in conformity with eqs. (3.101m), (3.101n) and (3.101 o). This, in fact, will help in fixing the other arbitrary functions as before.

(iv) Results in the Cartesian case: It is not difficult to derive the "potential" equations, similar to (3.114') and (3.115'), in the Cartesian coordinates. The results are pratically the same as (3.114') and (3.115') with the coefficient functions same as defined in eqs. (3.107) and (3.108) but now Z and $\bar{Z}$ are replced by x_2 and x_1 coordinates, respectively. For this purpose the other changes to be noted are (i) the replcement of $[\partial R / \partial \bar{Z}) - (\partial S / \partial Z)]$ in (3.115') by $[\partial R / \partial x_1) - (\partial S / \partial x_2)]$, and (ii) the revised definition of F, G_1, G_2, R and S, as

$$F = a_{112}\,(\partial V / \partial x_1) + a_{122}\,(\partial V / \partial x_2); G_1 = a_{122}\,(\partial V / \partial x_1)$$
$$+ a_{222}\,(\partial V / \partial x_2),$$
$$G_2 = a_{111}\,(\partial V / \partial x_1) + a_{112}(\partial V / \partial x_2); R = a_{11}(\partial V / \partial x_1)$$
$$+ a_{12}\,(\partial V / \partial x_2),$$
$$S = a_{22}\,(\partial V / \partial x_2) + a_{12}\,(\partial V / \partial x_1).$$

3.5 GENERALIZED ERMAKOV AND NEW ERMAKOV-TYPE SYSTEMS

3.5.1 Generalized Ermakov Systems

In Sect. 3.2.2, the applications of the Ermakov method, employed initially only to the simple TDHO problem, to more generalized systems, are already pointed out in terms of eqs. (3.20). As a special case, in this connection, while $x(t)$ satisfies (3.20a), $\rho(t)$ is found [72] to satisfy

$$\ddot{\rho} + \omega^2(t)\,\rho = (1/x\,\rho^2)\,\sum_i c_i\,(x/\rho)^{2m_i - 1},\ (i = 1, 2) \qquad (3.116)$$

where c_i and m_i are arbitrary constants. For $i = 1$ and 2, and with $m_1 = (m - 2)/2$, $m_2 = (2m\text{-}2)/2$, eq. (3.116) becomes

$$\ddot{\rho} + \omega^2(t)\,\rho = c_1\,x^{m-4}\,\rho^{1-m} + c_2\,x^{2m-4}\,\rho^{1-2m}. \qquad (3.117)$$

Among other generalizations of Ermakov systems considered by Reid and Ray [76], one is in relation with the nonlinear superposition law for the solutions of higher order nonlinear equations. We avoid the discussion of such generalizations here. It may be mentioned that all the above generalizations of Ermakov systems are essentially for 1D, TD systems, particularly for the 1D TDHO and $\rho(t)$ appears as an auxiliary variable needed to provide the invariant for the corresponding TD system in (1+1) dimensions. On the other hand, $\rho(t)$ also plays a specific role while looking for a physical interpretation of the derived invariant (cf. Chap. 7). In any case, it does not imply the generalization of Ermakov systems to higher space-dimension, in the present context (2+1) dimensions.

Other generalizations of Ermakov systems have recently been considered by Leach [87] and Athorne [88]. Leach finds an explanation for the nature of the Ermakov system described by

$$\ddot{x}_1 + \omega^2(t)\, x_1 = g_1(x_2 / x_1) / x_1^3\,, \tag{3.118a}$$

$$\ddot{x}_2 + \omega^2(t) x_2 = f_1(x_2 / x_1) / x_2^3, \tag{3.118b}$$

in terms of the symmetry algebra $s\ell(2, \mathcal{R})$. In this case, a transformation of time- and space-variables, namely $T = \cot(\int \rho^{-2}\, dt)$; $X = \rho^{-1}\, x_1 \csc T$; $Y = \rho^{-1}\, x_2 \csc T$, eliminates $\omega^2(t)$ from eqs. (3.118) and the newly introduced variable ρ is found to satisfy an equation of the type (3.6) with $k = 1$. No doubt, the generalized Ermakov systems are really Cartesian-forms of a system of equations, but for their deeper understanding from the symmetry considerations, the corresponding polar forms [87],

$$\ddot{r} - r\,\dot{\theta}^2 = F(\theta)/r^3;\; r\ddot{\theta} + 2\dot{r}\,\dot{\theta} = G(\theta)/r^3, \tag{3.119}$$

turn out to be more convenient. In fact, the symmetries explored, in this case, namely

$$G_1 = \partial / \partial t; G_2 = 2t(\partial / \partial t) + r(\partial / \partial r); G_3 = t^2(\partial / \partial t) + tr(\partial / \partial r), \tag{3.120}$$

correspond respectively to time-translation, self-similar and conformal transformations. Interestingly, if the system has Hamiltonian structure (the necessary condition for this is that $G = -(1/2)\,(\partial F / \partial \theta)$ in (3.119)) then the angular component of equations of motion (3.119) directly gives rise to the Ermakov invariant, i.e.

$$I = (1/2)\,(r^2\,\dot{\theta})^2 - \int G(\theta)\, d\theta = (1/2)\,[p_\theta^2 + F(\theta)]. \tag{3.121}$$

Not only this, the polar forms also suggest an easy way to generalize the Ermakov systems to higher dimensions. For example, in the 3D case, one can write [87] the equations of motion possessing the $s\ell(2, \mathcal{R})$ symmetry as

$$\ddot{r} - r\,\dot{\theta}^2 - r\,\sin^2\alpha\,\dot{\phi}^2 = F(\theta, \phi)/r^3,$$

$$r\,\ddot{\theta} + 2\dot{r}\,\dot{\theta} - r\,\dot{\phi}^2 \sin\theta\cos\theta = G(\theta,\phi)/r^3, \tag{3.122}$$

$$r\sin\theta\ddot{\phi} + 2\dot{r}\dot{\phi}\sin\theta + 2r\,\dot{\theta}\dot{\phi}\cos\theta = R(\theta,\phi)/r^3\sin\theta\,.$$

The system has a Hamiltonian provided

$$G(\theta,\phi) = -(1/2)\,(\partial F/\partial\theta),\; R(\theta,\phi) = -(1/2)\,(\partial F/\partial\phi),$$

This leads to the potential term $V = F(\theta, \phi)/2r^2$. The first invariant for this system was derived by Leach [87]. For further studies of 3D systems we refer to the recent work of Govinder et.al [90].

Athorne [88] makes use of the symmetry algebra of Leach and analyses a Kepler-Ermakov system of the type $\ddot{\vec{x}} + \omega^2(t)\,\vec{x} = v(\vec{x})r^3$ by setting $\omega^2(t) = 0$. Another interesting system studied in this context is that of coupled Pinney's equations [89]

$$\ddot{x}_1 + \omega^2(t)\,x_1 = \beta\,x_1^{-3} - \alpha\,x_1\,x_2^{-4}, \tag{3.123a}$$

$$\ddot{x}_2 + \omega^2(t)x_2 = \delta x_2^{-3} - \gamma\,x_2\,x_1^{-4}. \tag{3.123b}$$

In this case, however, there remains [88], in general, a difficulty of finding a Hamiltonian structure. In the absence of such a structure the system does not warrant much physical interest. For example, for the system (3.123) the Hamiltonian structure exists only for $\alpha = 3\delta$, $\gamma = 3\beta$ and with a kinetic term of hardly any physical interest. Inspite of the fact that the invariant associated with this system can be constructed explicitly, this system does not turn [88] out to be integrable even for $\omega(t)$ = const. i.e. even in a TID case. Depending upon the character of their superposition laws (i.e. on the nature of the functions f_1 and g_1 in eq. (3.118)) such as rational, algebraic or automorphic etc., the Ermakov systems can also be classified. For such details, we refer to the work of Athorne [90].

It is now clear that the Ermakov systems could be of both types—the one which have Hamiltonian structure and the other which do not have such a structure. From the recent group-theoretic studies of generalized coupled systems by Govinder and Leach [87], it appears that there can further be two separate situations—the one in which the system admits an invariant of angular momentum-type (i.e. Ermakov-type [74]) and the other in which the invariant is of energy integral-type (i.e. Lewis-type [63]). However, we restrict ourselves from such details here.

3.5.2 New Ermakov-type Systems

It may be mentioned that the variable $x_2(t)$ in (3.123) can, in principle, be regarded as $\rho(t)$ of eq. (3.6) or of (3.117) and is, in fact, a variable in the second space-dimension. On the other hand, the system studied by Leach [87] (cf. eq. (3.118)) no doubt reduces to the Ray and Reid form (3.20) on redefining g_1 and f_1 as $g_1(x_2/x_1) = (x_1/x_2)\, g(x_1/x_2)$; $f_1(x_2/x_1) = (x_2/x_1)\, f(x_2/x_1)$, but becomes structurally different from (3.20) in the sense that ρ-equation now appears as a by-product of the transformation. Thus, eqs. (3.118) and (3.123) describe certain TD coupled oscillators in 2D (with equal spring constants along each dimension) but without any auxiliary equation. It may be of interest to look for the Ermakov or Ermakov-type systems which remain integrable and also retain the Hamiltonian structure. In fact, the conditions under which the Ermakov system (3.20) is also Hamiltonian system are investigated by Cervero and Lejarreta [91] but by treating the auxiliary variable ρ at par with the space variable x. Recently, inspired by the results of dynamical algebraic approach (cf. Sect. 3.3.2), the author has proposed [92] a class of Ermakov-type systems in 2D which (*i*) deals with unequal but related spring constants, (*ii*) ensures, in general, a Hamiltonian structure, (*iii*) admits second order invariants, (*iv*) can involve fractional powers in the coupling terms, and (*v*) involves a pair of auxiliary equations in a natural manner. All these features could be important in characterizing the new Ermakov-type systems which, in fact, are capable of describing [92] TD, anharmonic and anisotropic oscillators in 2D.

Here, we just draw the attention to the systems (3.81) and (3.82) which are the special cases of the system described by the Lagrangian,

$$L = (1/2)\,(\dot{x}_1^2 + \dot{x}_2^2) - \alpha_1(t)\, x_1^2 - \alpha_2(t)\, x_2^2 - \beta_{10}\, x_1^{-2-n}\, x_2^n - \beta_{20}\, x_1^n\, x_2^{-2-n}, \tag{3.124}$$

with the corresponding pair of coupled equations of motion.

$$\begin{aligned} \ddot{x}_1 + 2\alpha_1(t)\, x_1 &= (n+2)\,\beta_{10}\, x_1^{-3-n}\, x_2^n - n\,\beta_{20}\, x_1^{n-1}\, x_2^{-2-n}, \\ \ddot{x}_2 + 2\alpha_2(t)\, x_2 &= -n\,\beta_{10}\, x_1^{-2-n}\, x_2^{n-1} + (n+2)\,\beta_{20}\, x_1^n\, x_2^{-3-n}. \end{aligned} \tag{3.124´}$$

The first invariant for this system is given by [92]

$$\begin{aligned} I = {} & (1/2)\,\psi(p_1^2 + p_2^2) + [(1/4)\,\ddot{\psi} + \alpha_1\psi]\, x_1^2 + \\ & [(1/4)\,\ddot{\psi} + \alpha_2\psi]\, x_2^2 + \beta_{10}\,\psi\, x_1^{-2-n}\, x_2^n + \beta_{20}\,\psi\, x_1^n\, x_2^{-2-n} \\ & - (1/2)\,\dot{\psi}\,(x_1 p_1 + x_2 p_2), \end{aligned} \tag{3.125}$$

where n is an arbitrary number; ψ is given as (cf. eqs. (3.76)) $\psi = c(\alpha_1 - \alpha_2)^{-1/2}$, and α_1 and α_2 are mutually related through (3.53), as before.

3.6 INTEGRABILITY OF NONCENTRAL TD SYSTEMS IN TWO DIMENSIONS

In the preceding Sections, we have constructed only one invariant for both 1D and 2D, TD systems. By constructing one invariant for 1D TD systems, no doubt, their integrability according to Whittaker's conjecture, is established but it is not so for the 2D, TD systems. In this latter case, particularly for the NC systems, there, however, remains a problem of obtaining the second invariant (I_2) which should not only be independent of the first one (I_1) (in the sense of the involuting property) but should also conform to the conditions (1.14) and (1.15) for the given Hamiltonian (H) of the system. As a matter of fact one should have

$$[I_1, H]_{PB} + (\partial I_1 / \partial t) = 0; [I_2, H]_{PB} + (\partial I_2 / \partial t) = 0; [I_1, I_2]_{PB} = 0, \tag{3.126a, b, c}$$

where H is now given by

$$H = (1/2)(p_1^2 + p_2^2) + V(x_1, x_2, t). \tag{3.127}$$

For this purpose, all the methods discussed in Sect. 3.3 in terms of the simple polynomial ansatz start exhibiting the complexity from the point of view of constructing the higher order invariants, but somehow do not even ensure the existence of I_2. In view of this complexity although it becomes difficult to question the viability of these methods as far as the construction of I_2 is concerned, but the fact is that there does not exist at present a general criterion which can guarantee the integrability of the higher-dimensional TD systems, may be in terms of nonpolynomial form of the invariant. In what follows, we make an attempt to set some further restrictions on the coefficient functions in the polynomial ansatz by assuming the existence of the second invariant I_2 for TD systems in 2D.

It is true that the expressions obtained for the coefficient functions, in general, involve several arbitrary functions (of time)/constants which are normally set in accordance with the parameters of the potential function V. Very often these arbitrary functions/constants outnumber the parameters of the system and a lot of freedom is left to fix at least a few of them. Therefore, the spirit in the following derivation is to impose some further restrictions on the coefficient functions in veiw of (3.126) alongwith the ones already required for the derivation of a particular invariant using the rationalization method. The results are given only for a few cases differing mainly due to the different order of I_1 and I_2, in particular we consider the cases (L,L), (L,Q), (L,C), (Q,Q), (Q,C) and (C,C) where L, Q, C respectively stand for linear, quadratic, cubic nature of (I_1,I_2). We first give results for the (C,C) case and subsequently obtain the results for other cases by setting some of the coefficient functions to zero.

(i) (C, C) case: For the Hamiltonian (3.127) and the ansatz

$$I_1 = a_0 + a_i \xi_i + (1/2)\, a_{ij} \xi_i \xi_j + (1/6)\, a_{ijk} \xi_i \xi_j \xi_k, \quad (3.128)$$

$$I_2 = b_0 + b_i \xi_i + (1/2)\, b_{ij} \xi_i \xi_j + (1/6)\, b_{ijk} \xi_i \xi_j \xi_k, \quad (3.129)$$

while the requirements (3.126a) and (3.126b) lead to the same set of PDE's for both the functions a's and b's as (3.96) to (3.100), the use of (3.126c), in the spirit of (1.17), will give rise to the restrictions on them as

$$a_{ijk,m}\, b_{\ell nm} - b_{ijk,m}\, a_{\ell nm} = 0, \quad (3.130)$$

$$a_{ijk,m}\, b_{\ell m} + 3a_{ij,m}\, b_{k\ell m} - b_{ijk,m}\, a_{\ell m} - 3b_{ij,m}\, a_{k\ell m} = 0, \quad (3.131)$$

$$a_{ijk,m}\, b_m + 3a_{ij,m}\, b_{km} - 6a_{i,m}\, b_{kjm} - b_{ijk,m}\, a_m$$
$$-3b_{ij,m}\, a_{km} - 6b_{i,m}\, a_{kjm} = 0, \quad (3.132)$$

$$a_{ij,m}\, b_m + 2a_{i,m}\, b_{jm} + 2a_{o,m}\, b_{ijm} - b_{ij,m}\, a_m$$
$$-2b_{i,m}\, a_{jm} - 2b_{o,m}\, a_{ijm} = 0, \quad (3.133)$$

$$a_{i,m}\, b_m + a_{o,m}\, b_{im} - b_{i,m}\, a_m - b_{o,m}\, a_{im} = 0, \quad (3.134)$$

$$a_{o,m}\, b_m - b_{o,m}\, a_m = 0. \quad (3.135)$$

Thus, some of the arbitrary functions/constants left undetermined in the expressions for a's in Sect. 3.4.2 (and now in the similar expression for b's) can be fixed with the help of above restrictions on a's and b's and there is a possibility of getting an involutively independent second invariant I_2 for the system (3.127).

(ii) (Q,C) case: By setting $a_{ijk} = 0$ all throughout in eqs. (3.128) to (3.135), one can obtain the conditions on a's and b's. In fact, in this case, while eq. (3.130) does not appear and eqs. (3.134) and (3.135) remain the same, eqs. (3.131) to (3.133) respectively take the form

$$3a_{ij,m}\, b_{k\ell m} - b_{ijk,m}\, a_{\ell m} = 0,$$

$$3a_{ij,m}\, b_{km} + 6a_{i,m}\, b_{kjm} - b_{ijk,m}\, a_m - 3b_{ij,m}\, a_{km} = 0,$$

$$a_{ij,m}\, b_m + 2a_{i,m}\, b_{jm} + 2a_{o,m}\, b_{ijm} - b_{ij,m}\, a_m - 2b_{i,m}\, a_{jm} = 0.$$

(iii) (L,C) case: Set a_{ijk} and a_{ij} to zero in eqs. (3.128) to (3.135). As a result, eqs. (3.130) and (3.131) are absent in this case and eq. (3.135) remains unchanged, leaving eqs. (3.132) to (3.134), respectively in the form

$$6a_{i,m}\ b_{kjm} - b_{ijk,m}\ a_m = 0;\ 2a_{i,m}\ b_{jm} + 2a_{o,m}\ b_{ijm} - a_{ij,m}\ a_m = 0,$$

$$a_{i,m}\ b_m + a_{o,m}\ b_{im} - b_{i,m}\ a_m = 0.$$

(iv) (Q,Q) case: Set $a_{ijk} = b_{ijk} = 0$ in eqs. (3.128) to (3.135). Consequently, eqs. (3.130) and (3.131) now do not appear and eqs. (3.134) and (3.135) remain the same. Eqs. (3.132) and (3.133), in this case, respectively take the form

$$a_{ij,m}\ b_{km} - b_{ij,m}\ a_{km} = 0,$$

$$a_{ij,m}\ b_m + 2a_{i,m}\ b_{jm} + b_{ij,m}\ a_m - 2b_{i,m}\ a_{jm} = 0.$$

(v) (L, Q) case: Set $a_{ijk} = a_{ij} = b_{ijk} = 0$ in eqs. (3.128) to (3.135). In this case, eqs. (3.130) to (3.132) are absent and eq. (3.135) remains unaffected. Eqs. (3.133) and (3.134), respectively now reduce to the form

$$2a_{i,m}\ b_{jm} - b_{ij,m}\ a_m = 0;\ a_{i,m}\ b_m + a_{o,m}\ b_{im} - b_{i,m}\ a_m = 0.$$

(vi) (L,L) case: For this simplest case, one sets $a_{ijk} = a_{ij} = b_{ijk} = b_{ij} = 0$, in eqs. (3.128) to (3.135), as before and obtains (3.134) in the form $a_{i,m}\ b_m - b_{i,m}\ a_m = 0$, in addition to eq. (3.135). All eqs. (3.130) to (3.133), in this case, are absent.

3.7 CONCLUDING DISCUSSION

For the purposes of constructing exact invariants for the systems involving explicit time-dependence, a brief survey of various methods used in the literature is carried out. The number of methods used for 1D systems is still more than those used for 2D systems. In spite of the fact that it is easier to test the merit of a new method on a 1D system, not all the methods used for this purpose are yet applied to two- or higher-dimensional systems. In fact, the underlying intricacies of a method reflect much more when it is applied to higher dimensional systems. Sometimes the method also shows limitations when it is extended to higher dimensions.

In spite of so many methods for 1D systems, not many new systems are found for which the invariants can be constructed. In this respect very often the TDHO system with its possible generalizations has played a pivotal role. As a matter of fact while the TDHO has offered a testing ground for various methods, the TD anharmonic oscillator problem even in 1D could not be investigated in most of the methods except for the one used by Leach and his coworkers [69-71]. Some of the approaches (like the dynamical algebraic one or the transformation-group method, for that matter), no doubt, have an obvious capability from the point of view of extending them to the corresponding quantum domain but this would be without actually demonstrating their complete potential at the classical level. Although the inherent mathematical elegance in the dynamical algebraic approach has beautifully suggested the criterion for the relative time-dependence of various coupling terms in the 2D case (cf. Sect. 3.3.2), but it somehow does not help in providing the second invariant for these systems if it exists. There, however, remains the difficulty of closure of the algebra as one proceeds not only for obtaining the higher order invariants in this approach but also for accounting for the higher order anharmonic terms with positive powers in $V(x_1, x_2, t)$. For this purpose, it may be of interest to use the available [93] generalized versions of this method.

Whether it is 1D or 2D case, we have not looked into very general solutions of the derived potential equations (cf. eqs. (3.18), (3.45), (3.57) to (3.59) and (3.85)) in rigorous mathematical terms except for trying some of their particular solutions which are often assumed to be separable in coordinates and time variables. Although it is difficult to obtain such general solutions, but if one obtains them, then they may provide several new systems which have not been covered in this chapter. The potential equation (3.60), derived using complex coordinates for the 2D case, has a special status in the sense that it offers the invariants for TD central potentials of nonharmonic nature.

In view of the fact that one of the coefficient functions and subsequently the potential function $V(x,t)$ in the rationalization method in the 1D case, factorizes [73] in x- and t-variables in a rather natural manner (cf. eq. (3.86) for the coefficient function $b_1(x,t)$ and similar is the situation for $b_0(x,t)$ in the second order case), the role of the self-similar transformation (3.35d) in giving rise to the form (3.35g), however, remains questionable. Any way, the use of self-similar technique while suggests an easier method to derive the higher order invariants for the 1D case, it does not provide very different integrable systems as far as their functional forms are concerned.

A mention may be made of other functional forms of the invariants used for TD systems. In this Chapter, while our survey mainly concerns the polynomial (in momenta) forms, nonpolynomial forms for the TD systems, however, are not so frequently studied in the literature as for the TID systems. There has also been discussion on the complex invariants [94] for the TDHO system. However, such ideas are more relevant in the context of quantum mechanics. In fact, in this case one exploits the generalized canonical transformations of Lewis and Leach [95] to convert the TD system into a corresponding TID one and then looks for the invariants in the Heisenberg picture.

Goedert and Lewis [29] have recently studied the rational form of the invariant for 1D, TD systems using the momentum-resonance formulation of Lewis and Leach [29]. For the system (3.15), in this case one makes an ansatz for the invariant as

$$I(x,p,t) = c(x,t) + \sum_{n=1}^{N} \frac{v_n\ (x,\ t)}{p - u_n\ (x,t)}.$$

Here, I is a rational function of momentum with simple poles, which are called momentum-resonances. As before, the use of eqs. (1.14) and (1.15) leads to necessary and sufficient conditions on the functions c, v_n, u_n and subsequently one expresses the resonance type invariant, I, as a functional of the potential $V(x,t)$. The cases of one, two and three resonances corresponding, respectively to the values 1, 2, and 3, of N are investigaged. Although we have tried to obtain the conditions, in general, under which the second invariant for 2D, TD systems can be constructed in a polynomial form (cf. Sect. 3.6), but the fact is that the available methods are inadequate for this purpose. Perhaps, the second invariant for such systems could be of nonpolynomial form which, of course, have not been studied so far.

Finally, the important observation made already in connection with the studies of coupled oscillators, uniformly in both TD and TID cases, is concerning the powers of the coupling terms in the Lagrangian i.e. the terms of the type $x_1^m\ x_2^n$ (cf. eqs. (3.50) and (3.124)). Interestingly, it is found that this system admits an invariant only for the case when $m + n = -2$. While some of the integrable systems corresponding to this situation are listed in Table 2.1 (cf. Chap. 2), the time-dependence of the others is discussed in Sects. 3.3.1, 3.3.2 and 3.5 mainly in the context of generalized Ermakov systems. Although such a restriction on m and n appears in a natural manner in the rationalization method and also found to have a basis in the closure of dynamical algebra of phase space functions, but the question remains as to why only this peculiar restriction. On the other hand, the ad-hoc choices of m_1 and m_2, made by Ray and Reid [72] in their work long ago (cf. eq. (3.117)), also conform to such a restriction on m and n. From all this it appears that this restriction indeed has a much deeper origin than merely providing the integrable systems. It could well be that the special setting of the kinetic and harmonic terms in the potential in the Hamiltonian structure, is responsible for such a restriction.

4

Quantum Mechanics of Noncentral Time Independent Systems

4.1 INTRODUCTION

As mentioned in Chapter 1 there appear now several newly discovered phenomena in various branches of physics and chemistry whose theoretical understanding might require an account not only of harmonic NC but also of anharmonic central and NC potentials at the quantum level. Moreover, such techniques, if developed and used for the case of Schrodinger equation (SE) in this context, can as well find applications in other physical problems where the Schrodinger-like equations also appear [96]. As a matter of fact the problem of anharmonic (AH) oscillator, or more generally, that of polynomial interactions even in 1D in quantum theory continues to be an active area of research despite a long history (see, for example, Refs. [113] and [122]). This is mainly because these problems are, in general, not exactly solvable and most of the work during these years has been devoted to the development of various approximation techniques (see Sect. 4.5). Although these problems (both nonrelativistic and relativistic ones) could be of interest in their own right but, interestingly, as a model, they also guide to similar methods which could be of immediate use in the quantum field theory. What one actually carries out in the nonrelativistic case is nothing but a 1D quantum field theory.

The use of NC potentials poses [97] its own problems at the level of both classical and quantum mechanics. While the difficulties that arise at the quantum level will be discussed in Sect. 4.3, the ones at the classical level are well known in terms of the integrability (cf. Sect. 3.6) of the corresponding system. For example, the most general form of the Lagrangian for an NC harmonic potential (in the absence of damping) in 2D is given by

$$L = \frac{1}{2}(\dot{x}_1^2 + \dot{x}_2^2) - \frac{1}{2}(ax_1^2 + bx_2^2 + 2c\, x_1x_2),$$

which leads to the equations of motion

$$\ddot{x}_1 + ax_1 + cx_2 = 0, \quad \ddot{x}_2 + bx_2 + cx_1 = 0.$$

Note that the solution of these equations for a = b is trivial and for $a \neq b$, the solution can be obtained by a suitable scale transformation for x_1 and x_2. On the other hand, in the simplest case of an AH potential for which the Lagrangian can be written as

$$L = \frac{1}{2}(\dot{x}_1^2 + \dot{x}_2^2) - \frac{1}{2}(ax_1^2 + bx_2^2 + 2c\, x_1x_2 + dx_1^3 + ex_2^3$$

$$+ f\, x_1^2x_2 + g\, x_1x_2^2),$$

the corresponding equations of motion turn out to be nonlinear and coupled. The integrability of such equations, in general, has already been questioned in the literature. The problem becomes much more complicated as one incorporates higher order anharmonicities in the potential.

At the quantum level, however, the problem of nonlinearity does not arise. Unless the potential itself depends on the state function, the SE remains linear. In this context, although the quartic-type AH potential has been studied [98, 139] rather extensively in 1D but not many attempts have been made to study the SE with a potential having higher order anharmonicity in two and higher dimensions. Further, while a study of the SE in 3D becomes a difficult task, its solution even in 2D may be of interest in a number of physical problems, particularly in fibre optics [96] and quantum chemistry.

In this as well as in the next chapter our aim is to discuss and obtain exact closed-form solutions of the SE for a variety of central and NC potentials. If the exact solution of the SE even in a certain sector of the independent variable or for certain quantum states (say e.g. for the ground state) becomes available, it will be of practical importance for several reasons: (i) An unsolvable (in the sense of closed-form solutions) realistic problem may be approximated in a wide range of distance by an exactly solvable one if the wave function outside this range is very small. (ii) These exact (special) solutions may be useful to check the usual perturbative solutions for certain constraints on the potential parameters (coupling constants) or to make the perturbations around the exact solutions. Unfortunately, the catalogue of exactly solvable models is rather small. The often discussed exactly solvable systems in most of the textbooks are the harmonic oscillator and Coulomb potentials. In this and the next Chapter we shall try to go beyond these two well known systems.

In the next Section, we make a survey of quantum mechanics of central potentials with an emphasis on the methods used to search for the exactly solvable cases besides the standard examples of harmonic and Coulomb potentials. Difficulties in dealing with the NC potentials in this context, are demonstrated explicitly in Sect. 4.3. In Sect. 4.4, we list various methods used for obtaining the exact closed-form solution of the SE for a variety of NC potentials. Approximation methods used for handling the SE with NC potentials are breifly discussed in Sect. 4.5. In Sect. 4.6, the role of constraints on potential parameters vis-a-vis the exact solution of the SE is discussed. With reference to the exact solutions of the SE, the role of dynamical symmetries is also discussed in Sect. 4.7. Finally, concluding remarks are made in Sect. 4.8.

4.2 SURVEY OF QUANTUM MECHANICS OF CENTRAL POTENTIALS: EXACT SOLUTIONS OF SE

4.2.1 Radial Versus One-Dimensional SE

We consider the SE with a central potential $V(r)$ in D-dimensions. After defining the radial wavefunction $R(r)$ in the form $R(r) = r^{(1-D)/2}\phi(r)$, the reduced radial SE becomes ($\hbar = m = 1$)

$$\phi''(r) + B(r)\,\phi(r) = 0, \tag{4.1}$$

where

$$B(r) = 2E - 2V(r) - C(D,\wedge)/r^2 \tag{4.2}$$

with

$$C(D,\wedge) = \wedge(\wedge + D - 2) + (1/4)(D-1)(D-3).$$

Note that for $D = 3$ and $\wedge = \ell$, the centrifugal term in (4.2) attains its standard form [99] $\ell(\ell+1)/r^2$. Further, as far as the analytic structure of the reduced form of the SE for the potential $V(r)$ is concerned, it turns out to be the same as that of the 1D SE for the potential $V(x)$ except for the following points: (i) A centrifugal term $C(D,\wedge)/r^2$ appears in the potential which again for $D = 3$ and $\ell = 0$ can either be set equal to zero or else for $\ell \neq 0$ it just adds an inverse harmonic term [100] to $V(r)$. (ii) The domain over which the wavefunction $\phi(r)$ is defined, is $0 < r < \infty$; whereas in the 1D case the corresponding $\phi(x)$ is defined over the domain $-\infty < x < \infty$. It may be mentioned that by using the transformation, $z = \ln r$, the domain of

r in terms of the new variable z, can also be extended to $-\infty < z < \infty$, which, in fact, is the same as that of x.

Thus, to some extent the problem of central potentials in quantum mechanics can be studied in the same way as that of $1D$ potentials, particularly when one looks for exact closed-form solutions of the SE. As a matter of fact except for the difference in the domains of the independent variables and an additional $(1/r^2)$—term in the case of central potentials, the two systems can be studied using the identical mathematical techniques. This is, however, not the case with the NC potentials for which a separate study is necessary.

4.2.2 Methods for Central (or 1-D) Potentials

There have been a large number of attempts [100-126] to obtain an exact closed-form solution of the SE for a variety of 1D or central potentials. No doubt the solutions so obtained could be normalizable and well-behaved at the origin and at infinity (depending upon the type of the problem) but still the exactness of these solutions remains questionable in the sense that the methods used provide either only a few eigenvalues or else they become much involved as one proceeds to obtain the complete spectrum. For such reasons, some authors [108, 118, 123] call these solutions of the SE as partial or 'quasi exact' solutions. However, here we shall call these solutions to be exact even if they correspond to a few eigenvalues. The methods used can broadly be classified into four categories: (a) Eigenfunction-ansatz method, (b) Standard-differential-equation method, (c) Quasi-dynamical Lie algebraic method, and (d) Supersymmetric factorization method.

(*a*) *Eigenfunction-ansatz method:* This method, first adopted by Hautot [104] for a potential, $V(r) = Ar^2 + Br - D/r$, (with a constraint for values of the parameter D) and later by Singh *et al.* [109], Magyari [104], Flessas [101], De Souza Dutra [110], Kaushal [111, 112] and others [122], is based on an ansatz made for the solution of the SE for a given potential. The unknown parameters in the ansatz are obtained in terms of the parameters of the potential mainly by rationalizing the SE for this solution. Sometimes the number of potential parameters (couplings) outnumber the unknown parameters in the ansatz. As a result the solutions are obtained with certain constraints on the potential parameters. The study of these constraints again turns [117] out to be interesting from the point of view of studying the excited states. Further, this method also has an overlap with what is known [118] as Hill-determinant or continued-fractions method used for numerical determination of the eigenvalues of certain potentials. In fact, this latter method has also provided exact solutions of the SE when the condition for factorization of the infinite-dimensional Hill-determinant is satisfied. The normalizability aspect of the wavefunctions obtained in this latter method has been studied recently by Singh *et. al.*, [114] for the sextic AH potentials.

For $D = 3$, the SE (4.1) takes the form

$$\phi''(r) + [\lambda - v(r) - \frac{\ell(\ell+1)}{r^2}]\, \phi(r) = 0, \tag{4.3}$$

where $\lambda = 2mE/\hbar^2$, $v(r) = 2m\, V(r)/\hbar^2$. Now, one makes an ansatz for $\phi(r)$,

$$\phi(r) = f(r)\, \exp(g(r)). \tag{4.4}$$

The use of this form in (4.3) leads to

$$\phi''(r) = [g'' + g'^2 + (f'' + 2\, f'\, g')/f]\, \phi(r). \tag{4.5}$$

A comparison of (4.5) with (4.3) gives rise to

$$v(r) + \frac{\ell(\ell+1)}{r^2} - \lambda = g'' + g'^2 + (f'' + 2\, f'\, g')/f. \tag{4.6}$$

Note that for $f(r)$ = const., the last term on the right-hand side in (4.6) vanishes and some polynomial forms of $g(r)$ in r have provided the ground state solutions for a variety of polynomial potentials (see next subsection). On the other hand, if $f(r)$ is not a constant but again a polynomial in r, say, of the form $f_n(r) = \prod_i^n (r - \alpha_i^{(n)})$ for $n = 1, 2, \ldots$, and $f_0(r) = 1$ for $n = 0$, with $r = \alpha_i^{(n)}$ as the nodes of the wavefunction, then the ansatz (4.4) is also capable of providing the excited states for a number of power potentials including the singular ones. We shall return to some of the applications of this method in the next subsection. As far as the underlying mathematical intricacies in this method are concerned they have been well analyzed recently by Znojil and Leach [118].

(*b*) *Standard-differential-equation method:* In this method, the SE for the given potential is transformed into a standard differential equation by a suitable transformation of both the dependent and independent variables. The solutions of this transformed equation are already known in the mathematics literature. Very often differential equation satisfied by an orthgonal polynomial is obtained in this method. The cases of Coulomb and harmonic central potentials in 3D and harmonic oscillator potential in 1D are already known and discussed in several textbooks. For example, the solutions for these cases, respectively involve Laguerre and Hermite polynomials. Recently, this method has been used by Aly and Barut [113] to find exact solutions for three large classes of AH potentials. In particular, the transformations involving the Kummer equation, a class of Kamke's equation and a generalized hypergeometric equation are investigated.

In the past, several authors have used [125] general method of transformation of the SE into the hypergeometric equation. For specific purpose, however, one still has to find special transformations to identify the physical potential. Aly and Barut [113] have applied the variations of this technique to a variety of AH potentials and obtained the exact solutions for the following energy-dependent potentials:

(i) $V(x) = \alpha x - k\alpha x^2 - \frac{1}{4} a^2 x^4,$

(ii) $V(x) = \frac{3}{2} \alpha x^2 - k\alpha x^3 - \frac{1}{4} a^2 x^6,$

(iii) $V(x) = \frac{1}{2}\alpha(n-1)\ x^{n-2} - \frac{1}{4}\ \alpha^2 x^{2n-2} - k\alpha x^{n-1},$

(iv) $V(x) = -\beta\ kx + \left(\frac{3}{2}\ \alpha - \frac{1}{4}\ b^2\right)x^2 - k\alpha x^3 - \frac{1}{4}\alpha\beta\ x^4 - \frac{1}{4}\ \alpha^2 x^6.$

Here, the energy eigenvalue in the first three cases is $(-k^2)$ whereas in the last case it is $(\frac{1}{2}\ \beta - k^2)$.

(c) Quasidynamical Lie algebraic method: It has been known [107, 108, 123] that many exactly solvable problems may be related to a finite-dimensional representation of Lie algebra. In this context, Turbiner [108] used the well known representation of $sl(2, R)$ to treat the problem of sextic potential in the form:

$$J^+ = z^2 \frac{d}{dz} - 2jz,\ J^{\circ} = z\frac{d}{dz} - j,\ J^- = \frac{d}{dz}, \tag{4.7}$$

where $z = x^2$ and $j(j+1)$ is the eigenvalue of the Casimir operator. In fact, the commutators of these operators give the standard relations

$$[J^+, J^-] = -2J^{\circ},\ [J^{\circ}, J^{\pm}] = \pm\ J^{\pm}. \tag{4.8}$$

The main difficulty with this representation of the elements of $sl(2, R)$ is that here one is limited to a polynomial potential of degree six, namely the potential of the form,

$$V(x) = a^2 x^6 + 2ab\ x^4 + [b^4 - (4n+3)\ a]\ x^2, \tag{4.9}$$

has been the case of discussion of Turbiner and his coworkers [107, 108] while investigating the details of this method in mathematical terms. In (4.9), a and b are the real parameters with $a > 0$ and n is an integer. Recently, de Souza Dutra and Boschi Filho [123] used a generalization of this representation to attain polynomial potentials of higher degree via the algebra $so(2,1)$ which is isomorphic to $sl\ (2,R)$. The representation used is

$$T_1 = \alpha_2\ x^{2-j}\left(\frac{d}{dx}\right)^2 + \alpha_1\ x^{1-j}\left(\frac{d}{dx}\right) + \alpha_o\ x^{-j},$$

$$T_2 = -\frac{i}{j}\ x\left(\frac{d}{dx}\right) - i\sigma,\ T_3 = \alpha\, x^j,$$

which satisfies the $so(2,1)$ algebra, viz.,

$$[T_1, T_2] = -i\,T_1\,;\ [T_2, T_3] = -i\,T_3\,;[T_1, T_3] = -i\,T_2, \tag{4.10}$$

provided only that

$$\sigma = \frac{1}{2j}\left[\frac{\alpha_1}{\alpha_2} + j - 1\right],\ \lambda = -(2\,\alpha_2\ j^2)^{-1}.$$

Here the constants α_1, α_2 and j are selected to match the different elements of the Hamiltonian. Further, the work of de Souza Dutra and Boschi Filho provides a clue to expressing the most general form of the Hamiltonian in terms of the generators of the $sl(2, R)$ or $so(2,1)$ algebras. More recently Burrows et al. [124] have investigated the mathematical details of this method in terms of $so(3)$ or $so(2,1)$ algebras. It is suggested that the Hamiltonian when expressed as a bilinear function of the generators of these algebras admits only a part of the bound-state spectrum and the same can be calculated analytically to provide a finite set of (quasi) exact solutions.

(*d*) *Supersymmetric factorization method:* Supersymmetry-inspired factorization method [121] has also been used [119, 120] to obtain exact analytic solutions of the SE for one or a few quantum states of general polynomial and non-polynomial potentials. In this method, the Hamiltonian satisfying the SE

$$H\psi_n \equiv \left(\frac{-\hbar^2}{2m}\frac{d^2}{dx^2} + V_{eff}(x)\right)\psi_n(x) = E_n\,\psi_n\,(x), \tag{4.11}$$

for the n-th quantum state is written as [128]

$$H\_ = A^+\,A^- \tag{4.12}$$

such that the n-th eigenstate of this latter form of the Hamiltonian corresponds to eigen energy $E_n^{(-)} = 0$. In that case A^+ and A^- are given by

$$A^{\pm} = \pm \frac{\hbar}{\sqrt{2m}} \frac{d}{dx} + W(x), \tag{4.13}$$

where W(x) defined by

$$W(x) = -\frac{\hbar}{\sqrt{2m}} \frac{d}{dx} \ell n\, \psi_n^{(-)}(x), \tag{4.14}$$

is called the superpotential. From (4.12) and (4.13), the potential V_- corresponding to H_- is

$$V_- = W^2(x) - \frac{\hbar}{\sqrt{2m}} W'(x), \tag{4.15}$$

and the n-th eigenstate for this potential is obtained from (4.14) as

$$\psi_n^{(-)} = \exp\left(-\frac{\sqrt{2m}}{\hbar} \int^x W(x')\, dx' \right). \tag{4.16}$$

It may be noted that when n corresponds to the ground state, the relations (4.12) – (4.16) correspond to an unbroken supersymmetry and there exists a super-partner Hamiltonian $H_+ = A^- A^+$ corresponding to H_-. However, for the excited states such a correspondence may not occur. It is only H_- which is important here for the purposes of obtaining the exact solutions of (4.11). Now, if it turns out that V_{eff} in (4.11) and V_- in (4.15) differ only by a constant i.e. $V_- = V_{eff} + C$, one may write $E_n^{(-)} = E_n + C$. Since $E_n^{(-)} = 0$ by construction, the eigenvalue E_n is given by $E_n = -C$ and subsequently, the eigenfunction $\psi_n(x)$ corresponding to the n-th state becomes the same as $\psi_n^{(-)}(x)$. As far as the reproduction of the nodal structures of the wavefunction in this method is concerned, it can be done by choosing appropriate $W(x)$ which is consistent with (4.15). However, such a choice of $W(x)$ may not be unique [129]. One can always write another $W_1(x) = W(x) + f(x)$, which generates the same V_-.

Adhikari et al. have applied this method to study a generalized polynomial potential [119] of the form

$$V_{eff}(x) = \sum_{j=3}^{2N} b_{j-2}\, x^{j-2} + (\alpha / x) + \ell(\ell+1)\hbar^2 / (2mx^2), \; (b_{2N-2} > 0), \tag{4.17}$$

and a generalized nonpolynomial potential [130],

$$V_{eff}(x) = x^2 + \left(\sum_{j=0}^{m} \lambda_j \, x^{j-1} \right) \left(\sum_{j=0}^{m} g_j \, x^j \right)^{-1} + V_o \,, \qquad (4.18)$$

where V_0 is a constant.

In veiw of the recent studies by Beckers et al [130], a word of caution regarding the use of this method for central potentials is necessary. Beckers et al, have, in fact, shown that supersymmetric three-dimensional SE describing central problems does not lead, in general, to supersymmetric one-dimensional radial equation—a situation well supported by the work of Kostelecky and Nieto [130]. Further, within the framework of this method exactly solvable potentials can be understood in terms of a few basic ideas which include supersymmetric partner potentials, shape-invariance and operator transformations. For details we refer to the review by Cooper et al. [121].

4.2.3 Exactly Solvable Examples

Here we present only the results (i.e. the eigenvalues, eigenfunctions, constraints on the potential parameters if necessary to provide the exact solution of the SE, and a brief hint to the method used for this purpose) for a variety of potentials. In particular, the cases of polynomial, polynomial singular, nonpolynomial and exponential potentials are discussed. Although the first and second categories can be combined together (since the number of terms with negative power are not many), but we discuss them here separately.

(*a*) *Polynomial potentials:* In this case the most studied example is that of a sextic potential,

$$\mathrm{v}(r) = ar^2 - br^4 + cr^6, (a, b, c > 0). \qquad (4.19)$$

Using the eigenfunction-ansatz method the ground state energy from eq.(4.3) turns out to be [111]

$$\lambda_\ell = -\left(\ell + \frac{3}{2} \right) b \,/\, \sqrt{c}\,,$$

where b is fixed from the constraint

$$b = sgn(b)\; 2\sqrt{c}\; (3\sqrt{c} + 2\sqrt{c}\;\delta + 1)^{1/2}, \qquad (4.20)$$

with $\delta = (\ell + 1)$. The normalized eigenfunction is given by

$$\phi_\ell(r) = N_\ell \, r^\ell \, \exp[(br^2 - cr^4)/4\sqrt{c}], \tag{4.21}$$

with

$$N_\ell = \sqrt{2}\, c^{(2\ell+3)/16} \, \exp(-b^2/32\, c\sqrt{c}) \, [\Gamma((2\ell+3)/2).$$

$$D_{-(2\ell+3)/2}(-b/\sqrt{2}\, c^{3/4})]^{-1/2},$$

where $D_\nu(x)$ is the parabolic cylindrical function [127]. Using the same method the eigenvalues and the eigenfunction are also obtained for a even power potential of degree ten, viz.,

$$v(r) = ar^2 - br^4 + cr^6 - dr^8 + e\, r^{10}, \tag{4.22}$$

as

$$\lambda_\ell = -(2\ell + 3)\, \beta,$$

$$\phi_\ell(r) \sim r^{\ell+1} \exp\left(\frac{1}{2}\beta r^2 - \frac{1}{4}\alpha r^4 + \frac{1}{6}\tau r^6\right),$$

where

$$\beta = (4c\, e - d^2)/4e\sqrt{e}; \; \delta = \ell + 1,$$

and α, β τ, δ have to satisfy the following constraints:

$$\beta^2 - 2\alpha\delta - 3\alpha = a; \; 2\alpha\beta - 2\tau\delta - 5\tau = b.$$

(b) Polynomial potentials with singular terms: In many physical problems (particularly in atomic and molecular physics and quark dynamics), the appearance of a Coulomb term alongwith the positive power terms in $v(r)$ becomes an asset. The eigenfunction-ansatz-method has also been used by several authors [110, 112, 122] to study such potentials. The exact ground state solution of the SE for the potential (cf. form (4.17))

$$v(r) = -\frac{a_o}{r} + \sum_{k=1}^{n} a_k r^k, \tag{4.23}$$

is obtained by de Souza Dutra [110] with a view to accommodating some odd power terms in (4.23). However, the transformation, $r = u^2$ used by de Souza Dutra reduces the problem again to a study of even power potentials. On the other hand, the solution for (4.23) with all terms up to $n \leq 8$ is also obtained [112] by using the same method. In general, it is found that the eigenvalues and eigenfunctions for any n in (4.23) do not depend on all a_k's; rather if we write $n = 2N$, then a_k's of interest are a_n, a_{n-1} for $N = 2$, a_n,

a_{n-1}, a_{n-2} for $N = 3$, a_n, a_{n-1}, a_{n-2}, a_{n-3} for $N = 4$, and so on. Further, even and double even power (which respectively correspond to $2k$ and $(2k+2)$-type powers in (4.23), where k is an odd integer) potentials are found [112] to exhibit somewhat distinct features as far as the corresponding eigenvalues are concerned.

For $n = 2M$, the form (4.23) is also investigated recently by Simsek and Ozcelik [122] with a view to obtaining a complete spectrum of eigenvalues. They consider (4.23) as a special case of a general Laurent-type potential of the form

$$v(r) = \sum_{i=-2j}^{2M} a_i\, r^i, \; (j,\, M > 0). \tag{4.24}$$

This Laurent-type potential in the reduced form (4.23) has a variety of applications in various branches of physics. Simsek and Ozcelik choose $f(r)$ and $g(r)$ in (4.4) as

$$f_n(r) = \prod_{i=1}^{n} (r - \alpha_i^{(n)}), \; n = 1, 2, \ldots, \; f_o(r) = 1,$$

$$g(r) = \sum_{k=0}^{M} \frac{\beta_k\, r^{k+1}}{(k+1)} + \delta\, \ell n\, r,$$

and obtain the energy spectrum in the form

$$E_n = -\left[\beta_o^2 + \beta_1(2\delta + 2n + 1) + 2\sum_{k=2}^{M} \beta_k \left(\sum_{i=1}^{n} (\alpha_i^{(n)})^{k-1}\right)\right],$$

with the relations among the eigenfunction parameters $\alpha_i^{(n)}$, β_k and δ (first three sets) as

$$2n\,\delta + n(n-1) + 2\sum_{k=0}^{M} \beta_k \left(\sum_{i=1}^{n} (\alpha_i^{(n)})^{k+1}\right) = 0,$$

$$[(2n-2)\,\delta + n(n-3) + 2] \sum_{i=1}^{n} \alpha_i^{(n)} + 2\sum_{k=1}^{M} \beta_k \left(\sum_{i \neq j}^{n} \alpha_i^{(n)} (\alpha_j^{(n)})^{k+1}\right) = 0,$$

$$[(2n-4)\,\delta + n(n-5) + 6] \sum_{i<j}^{n} \alpha_i^{(n)}\, \alpha_j^{(n)}$$

$$+2 \sum_{k=0}^{M} \beta_k \left(\sum_{i<j<m}^{n} \alpha_i^{(n)} \; \alpha_j^{(n)} (\alpha_m^{(n)})^{k+1} \right) = 0.$$

These results, in general, are found in agreement with those obtained by other methods such as the supersymmetric factorization method [119].

Another singular power potential, which has attracted [97,115-117] a great deal of interest in recent years, is

$$v(r) = ar^2 + br^{-4} + cr^{-6}, \; (a > 0, \; c > 0). \tag{4.25}$$

For this potential, the g.s. energy and the corresponding eigenfunction are found to be [115]

$$E_o = \sqrt{a} \; (4 + b\sqrt{c}),$$

$$\phi_o(r) = N_o \; r^{(3+b/\sqrt{c})/2} \exp\left[-\frac{1}{2} \left(\sqrt{a} r^2 + \sqrt{c} \; r^{-2} \right) \right], \tag{4.26}$$

alongwith the constraint

$$(2\sqrt{c} + b)^2 = c \; [(2\ell+1)^2 + 8\sqrt{ac}], \tag{4.26$'$}$$

on the potential parameters. First few excited states for (4.25) are also investigated [115-117] within this framework.

(c) Non-polynomial potentials: The only nonpolynomial potential which has been of great interest in the literature is of the form

$$v(r) = r^2 + \lambda r^2/(1 + g \; r^2). \tag{4.27}$$

In fact, this potential has been a testing ground in the past for the relative merit of different approximation methods. Bose and Varma[126] have made an attempt to obtain a class of exact solutions of the SE for this potential using the well-known series-expansion-ansatz method. In spite of the fact that this method is capable of providing the complete spectrum of the problem, in this case, however only a few bound states are derived and that too with a large number of constraints on the potential parameters. The problem remains more or less the same when the elegance of the supersymmetric factorization method is demonstrated for the m-space dimension generalized version (4.18) of (4.27) by Adhikari et al. [119].

(d) Exponential potentials: The eigenfunction-ansatz method has been used successfully to obtain an exact closed-form solution of the SE for

the well known Morse potential

$$v(r) = v_o \; [e^{-2a(r-r_o)} - 2e^{-a(r-r_o)}]. \tag{4.28}$$

Interestingly, without any constraint on the potential parameter, the eigenvalues and eigenfunctions for (4.28) turn out to be [97]

$$\lambda = -\left(\frac{1}{2}\, a - \sqrt{v_o}\right)^2 ,$$

$$\phi_o(u) = N \; \exp\left[\left(\frac{1}{2}\, a - \sqrt{v_o}\right) u - (\sqrt{v_o}\, / a) \; \exp(-au)\right],$$

where $u = r - r_0$. Note that for the Morse central potential the solution $\phi_0(u)$ obtained here also does not vanish at $r = 0$, like other solutions given in the literature. Using the same method recently Simsek and Ozcelik [122] have studied a four parameter exponential potential of the form

$$v(r) = \frac{A\, e^{ar}}{(C + e^{ar})} - \frac{B\, e^{ar}}{(C + e^{ar})^2} \, , \; (A, \; B > 0). \tag{4.29}$$

It may be noted that for different choices of the parameters the form (4.29) reduces to several standard potentials which are frequently used in various branches of physics and molecular chemistry. For example, the following table depicts some such standard cases:

1. $C = 1$	(Rosen-Morse form)
2. $B = 0$, $C = -1$, a = negative	(Hulthen form)
3. $C = -1$, a = negative	(Hulthen-like effective form)
4. $C = -1$, a = negative with (ar) small	(screened-Coulomb effective form)

Further, somewhat complicated expressions are obtained for the eigenvalues and eigenfunctions for the potential (4.29).

4.3 DIFFICULTIES IN DEALING WITH NC POTENTIALS

Difficulties in dealing with the NC potentials at the classical level, as mentioned in Sect. 4.1, are of different nature from those of dealing with them at the quantum level. Here, we briefly point out such difficulties in the quantum context, particularly when one tries to obtain an exact solution of the SE for these potentials. It may be mentioned that besides limiting the existence and the validity of the exact solution so obtained, the constraints on the potential parameters also create the problem at the stage of computing the excited states for both central [117] and NC potentials. Here, however, we emphasize other type of difficulties for the NC case within the framework of the eigenfunction-ansatz method.

The form of the SE to be investigated in this case is

$$\partial^2\phi/\partial x_1^2 + \partial^2\phi/\partial x_2^2 + [\lambda - v(x_1, x_2)]\ \phi(x_1, x_2) = 0, \quad (4.30)$$

where $\lambda = 2mE/\hbar^2$, $v(x_1, x_2) = 2m\ V(x_1, x_2)/\hbar^2$. As it was done for the central potentials, an ansatz for the eigenfunction can be made as

$$\phi(x_1, x_2) = N\ f(x_1, x_2)\ \exp(g(x_1, x_2)), \quad (4.31)$$

and use this form in (4.30) to get the expression

$$v(x_1, x_2) - \lambda = g_{11} + g_1^2 + g_{22} + g_2^2 + [(f_{11} + 2f_1 g_1 + f_{22} + 2f_2 g_2)/f], \quad (4.32)$$

in analogy with the expression (4.6). Here the subscripts on the functions of two variables like g, f and on V (in future) are in the spirit of partial derivatives w.r.t. x_1 or x_2 or both, e.g. $g_{ij} = \partial^2 g/\partial x_i \partial x_j$. Now the rationalization of this expression for a given NC potential suggests the determination not only of the unknown functions $g(x_1,x_2)$ and $f(x_1,x_2)$ but also of the eigenvalue λ, sometimes leaving behind certain constraints (which mainly depend on the nature of the potential) on the potential parameters. Naturally, the number of equations to be dealt with in this case is much more than that for the central potentials. Note that for $f(x_1,x_2)$ = const. in (4.31), the last term in (4.32) vanishes and this greatly simplifies the problem and more so for the potentials separable in x_1 and x_2 variables. This will, however, provide only the ground state solution of the SE (4.30). In fact, for the excited states an account of $f(x_1,x_2)$, mainly of a polynomial form either in one or in both the variables x_1 and x_2, is necessary, say the form

$$f(x_1, x_2) \equiv f_{n,m}(x_1, x_2) = \prod_{i=1}^{n}(x_1 - \alpha_i^{(n)}) \prod_{j=1}^{m}(x_2 - \beta_i^{(m)}), \text{ for } n, m = 1,2,$$

....., could be a convenient one. At this stage the difficulty may also arise when $f(x_1,x_2)$ (or for that matter $g(x_1,x_2)$) is not separable in x_1, and x_2.

Another stage at which the difficutly in handling the NC potentials may arise is during the normalization of the eigenfunction (4.31). In particular, the normalization integral of $\phi(x_1,x_2)$, given by

$$\int_{-\infty}^{+\infty}\int_{-\infty}^{+\infty} |\phi(x_1, x_2)|^2\ dx_1\ dx_2 = 1, \quad (4.33)$$

turns out to be an improper integral for certain forms of $V(x_1,x_2)$. As a result, it is not in general possible to obtain nonzero eigenvalues with normalizable eigenfunctions for these forms of $V(x_1,x_2)$. Now we discuss some explicit forms of $V(x_1,x_2)$ for which such difficulties are transparent.

Consider, for example, the solution of (4.30) for an anharmonic quartic potential [97] of the form

$$V(x_1,x_2) = b_{20}\, x_1^2 + b_{02}\, x_2^2 + b_{22}\, x_1^2\, x_2^2 + b_{40}\, x_1^4 + b_{04}\, x_2^4. \quad (4.34)$$

The use of the ansatz (4.31) with $f(x_1,x_2) = 1$ and

$$g(x_1,x_2) = \beta_{10}x_1 + \beta_{01}x_2 + \beta_{20}x_1^2 + \beta_{02}x_2^2 + \beta_{11}\, x_1x_2 + \beta_{21}\, x_1^2x_2$$

$$+ \beta_{12}\, x_1x_2^2 + \beta_{30}\, x_1^3 + \beta_{03}\, x_2^3, \quad (4.35)$$

in (4.32) after rationalization leads to a number of equations between β_{ij}'s and b_{ij}'s alongwith the one for the eigenvalue as

$$\lambda = -(2\beta_{20} + 2\beta_{02} + \beta_{10}^2 + \beta_{01}^2). \quad (4.36)$$

An analysis of this over-determined system of equations reveals [97] that

$$\beta_{01} = \pm\, i\beta_{10}\, ;\; \beta_{11} = \pm\, 2i\beta_{20} = \mp\, 2i\beta_{02},$$

$$\beta_{21} = -3\beta_{03}\, ;\; \beta_{12} = -3\beta_{30};\; \beta_{21} = \pm\, i\beta_{12}. \quad (4.37)$$

It can be seen that for these values of β_{ij}'s not only the λ (from (4.36)) becomes zero but the normalization integral also (in (4.33)) turns out to be an improper one. In fact, similar is the case if we consider the potential

$$V(x_1,x_2) = b_{10}\, x_1 + b_{01}\, x_2 + b_{20}\, x_1^2 + b_{02}\, x_2^2 +$$

$$b_{22}\, x_1^2x_2^2 + b_{40}\, x_1^4 + b_{04}\, x_2^4. \quad (4.38)$$

It is, however, found that if $V(x_1,x_2)$ involves inverse harmonic terms like $(b_1 / x_1^2 + b_2 / x_2^2)$ and/or crossed terms like $(b_3\, x_1/x_2 + b_4\, x_2/x_1)$, ... , then some of these potentials admit exact solutions of the SE. We shall return to these cases in the next section.

Another problem which also arises even at the classical level with NC potentials is the one related to finding a unique absolute minimum of $V(x_1,x_2)$ in the sense that it conforms to both x_1 and x_2 variables, particularly when $V(x_1,x_2)$ is not separable. In the absence of such a situation, it is rather difficult to obtain a complete spectrum of distinct eigenvalues of the system. In fact, to obtain the extremum of a function of two variables, $V(x_1,x_2)$, there exists the restriction not only on V_{11} and V_{22} but also on the crossed derivative V_{12} (or V_{21}) which more often conforms to a saddle point than to an extremum point.

4.4 QUANTUM MECHANICS OF NC POTENTIALS: EXACT SOLUTIONS OF THE SE

4.4.1 Various Methods

Inspite of the importance of NC forces in various physical problems not many methods have been developed so far to obtain exact solution of the SE. Available methods can broadly be classified into three categories: (a) Method of separation of variables, (b) Eigenfunction - ansatz method and (c) Standard - differential - equation method. Here, we shall refrain from giving the details of the first method as it is already discussed in several textbooks but rather demonstrate the applications of the other two methods which appear to carry the spirit of nonseparability of potentials in terms of its variables.

(*a*) *Method of separation of variables:* Using this well known method in polar coordinates Khare and Bhaduri [131] have recently obtained exact solutions of the SE for several nontrivial but separable NC potentials. They have shown that those NC potentials for which the SE is separable are analytically solvable, provided the separated problem for each of the variables belongs to the class of exactly solvable one-dimensional problem. In particular, the classes of potentials investigatged are $V(r, \theta) = U(r) + U_1(\theta)/r^2$, and $V(r, \theta, \varphi) = U(r) + U_1(\theta)/r^2 + U_2(\varphi)/r^2 \sin^2\theta$, in two and three dimensions, respectively. In 2D they have investigated a general case when $U(r) = \frac{1}{4}\omega^2 r^2$ and $U_1(\theta) = G/\sin^2 p\theta$. For $p = 3$, an interesting application of these results is further demonstrated to the well known but thus far difficult case of Calogero system [132] described by the Hamiltonian

$$H = -\sum_i (\partial^2/\partial x_i^2) + \frac{1}{12}\omega^2 \sum_{i<j}(x_i - x_j)^2 + \frac{1}{2}\gamma \sum_{i<j}(x_i - x_j)^{-2}. \tag{4.39}$$

In fact, for the three particle case this system, after the centre of mass motion is factored out, involves only two degrees of freedom which may be mapped on to the (r, θ) coordinates of a particle in a NC potential. It may be of interest to compare qualitatively the exact solutions for the Calogero potential in (4.39) and for the harmonic plus inverse harmonic potential in 1D. In the latter case, the coupling of the inverse harmonic term is found [100] to attain only certain discrete values as a result of the normalization of the eigenfunction.

(b) Eigenfunction-ansatz method: This method, already described in Sect. 4.2 for the central potentials and outlined in Sect. 4.3 for the NC potentials, has been extensively investigated by Taylor and Leach [133] for polynomial potentials of anharmonic nature. In fact, this method in its preliminary form was also used by Makarewicz [134] to study a NC sextic potential. Taylor and Leach [133] follow a general procedure for the construction of closed-form solutions of (4.30) for a class of NC potentials involving quartic- and/or sextic-type anharmonicity. They look for the examples in which the last term in the expression (4.32) remains a polynomial in x_1 and x_2 variables for a given form of $g(x_1,x_2)$. We shall return to some further applications of this method in the next subsection.

(c) Standard-differential equation method: In this category we list the method of Guha and Mukherjee [135] who have studied a NC parabolic potential. The same will be discussed in the next subsection. In fact, the SE for this potential finally reduces to a Laguerre equation after performing a sequence of transformations.

(d) Other methods: Among other methods we briefly mention here about the complexification method adopted by Turbiner and Ushveridze [136]. They use the so-called conformal transformation $\xi = f(z)$ where $\xi = u + iw$, $z = x_1 + ix_2$, to convert the SE (4.30) into the form

$$\{-(\partial^2/\partial u^2 + \partial^2/\partial w^2) + (\mathcal{V} - E)/F\}\, \phi\, (u,w) = 0 \qquad (4.40)$$

where $\mathcal{V}$ is identified as $\mathcal{V} = F\,[A(u) + B(w)]$, with $F = \left|f'\,(z)\right|^2$ for the case when $E = 0$. Here $A(u)$ and $B(w)$ are the arbitrary functions of the corresponding arguments. Thus, for the nontrivial choice $f(z) = z^2$ which implies $u = x_1^2 - x_2^2, w = 2x_1x_2$, and $A(u) = (\frac{1}{4}u^2 - \alpha)$, $B(w) = (1 + a)^2 w^2 - \beta$, the exact solution of (4.40) is obtained for the potential

$$\begin{aligned} V(x_1, x_2) &= x_1^6 + x_2^6 + a(a+2)\, x_1^2\, x_2^2\, (x_1^2 + x_2^2) + \\ &\quad [a + 3 - \{\tfrac{1}{2}\,(1+a)n + m\}]\,(x_1^2 + x_2^2), \end{aligned}$$

as

$$\phi_{n,m}\,(x_1, x_2) = H_n(\sqrt{(1+a)}\; x_1x_2).\; H_m\left(\frac{x_1^2 - x_2^2}{\sqrt{2}}\right)$$

$$\cdot \exp[-\frac{1}{4}\,(x_1^4 + x_2^4 + 2ax_1^2\, x_2^2)]\,, \qquad (4.41)$$

where H_k are Hermite polynomials, n and m are integers and a is the asymmetry parameter. For this rather simple choice the degeneracy of energy levels and the behaviour of node surfaces of the wave function (4.41) are studied. However the study of the case with nonzero eigenvalues turns out to be difficult in this method.

Next we discuss some exactly solvable examples within the framework of the above methods.

4.4.2 Some Exactly Solvable Examples

(*a*) *Power potentials:* Here we list only the results for the ground state of four NC potentials obtained [97] by using the eigenfunction - ansatz method. Two of them correspond to harmonic and the other two correspond to anharmonic (quartic) NC potentials. Also note that here $a_i = 2m\, b_i/\hbar^2$, $a_{ij} = 2m\, b_{ij}/\hbar^2$ are used.

(*i*) For the potential

$$V(x_1, x_2) = b_{20}\, x_1^2 + b_{02}\, x_2^2 + (b_1 / x_1^2) + (b_2 / x_2^2), \tag{4.42}$$

the eigenvalue and the eigenfunction are obtained as

$$\lambda = \sqrt{a_{20}}\ \{2 + (1 + 4a_1)^{1/2}\} + \sqrt{a_{02}}\ \{2 + (1 + 4a_2)^{1/2}\},$$

$$\phi(x_1, x_2) = N.\,(x_1 x_2)^{1/2}\, x_1^{\{(1+4a_1)^{1/2}/2\}}\, x_2^{\{(1+4a_2)^{1/2}/2\}}$$

$$.\exp(-(\sqrt{a_{20}}\ x_1^2 + \sqrt{a_{02}}\ x_2^2)/2),$$

where the normalization constant N can be obtained from (4.33).

(*ii*) For the potential

$$V(x_1, x_2) = b_{20}\, x_1^2 + b_{02}\, x_2^2 + b_{11} x_1 x_2 + b_3 (x_1 / x_2) + b_4 (x_2 / x_1), \tag{4.43}$$

the eigenvalue and eigenfunction are given by

$$\lambda = 3[(a_{20} - \bar{a}^2)^{1/2} + (a_{02} - \bar{a}^2)^{1/2}],$$

$$\phi(x_1, x_2) = N.(x_1, x_2) \exp[-\{(a_{20} - \bar{a}^2)^{1/2}\, x_1^2 + (a_{02} - \bar{a}^2)^{1/2}\, x_2^2 - 2\bar{a}\, x_1 x_2\}/2].$$

with the constraint,

$$a_{11} = -2\bar{a}\,[(a_{20} - \bar{a}^2)^{1/2} + (a_{02} - \bar{a}^2)^{1/2}],$$

on the potential parameters. Here, $a_3 = a_4 = \bar{a}$ and N can be obtained again from (4.33).

(*iii*) For the potential

$$V(x_1, x_2) = b_{10}\, x_1 + b_{01}\, x_2 + b_{20}\, x_1^2 + b_{02}\, x_2^2 + b_{30}\, x_1^3 + b_{03}\, x_2^3 + b_{40}\, x_1^4 + b_{04}\, x_2^4 + (b_1 / x_1^2) + (b_2 / x_2^2), \tag{4.44}$$

the eigenvalue and the eigenfunction are found as

$$\lambda = -[\sqrt{a_{20}}\ \{2 + (1 + 4a_1)^{1/2}\} + \sqrt{a_{02}}\ \{2 + (1 + 4a_2)^{1/2}\}],$$

$$\phi(x_1, x_2) = N.\,(x_1 x_2)^{1/2}\, x_1^{\{(1+4a_1)^{1/2}/2\}}\, x_2^{\{(1+4a_2)^{1/2}/2\}}\, \exp[-\{3(\sqrt{a_{20}}\ x_1^2 + \sqrt{a_{02}}\ x_2^2) + 2\,(\sqrt{a_{40}}\ x_1^3 + \sqrt{a_{04}}\ x_2^3)\}/6],$$

with the following four constraints on the potential parameters:

$$a_{30} = (2a_{20}\, a_{40})^{1/2}\ ;\ a_{03} = (2a_{02}\, a_{04})^{1/2},$$

$$a_{10} = \sqrt{a_{40}}\ [3 \pm (1 + 4a_1)^{1/2}]\ ;\ a_{01} = \sqrt{a_{04}}\ [3 \pm (1 + 4a_2)^{1/2}].$$

(*iv*) Finally for the simplified potential (in the sense of the number of parameters)

$$V(x_1, x_2) = 3(b_6\, x_1 + b_5 x_2) + b_{20} x_1^2 + b_{02}\, x_2^2 + b_{11}\, x_1 x_2 + \frac{1}{8}(b_5^2\, x_1^4\, b_6^2\, x_2^4) + \frac{1}{4}\bar{b}\,(b_5 x_1^3 + b_6\, x_2^3) + \frac{1}{2}(b_5^2 + b_6^2) x_1^2 x_2^2 + \frac{1}{2} b_5 b_6 (x_1\, x_2^3 + x_1^3 x_2) + \bar{b}\, [(x_1 / x_2) + (x_2 / x_1)] + b_5 (x_1^2 / x_2) + b_6 (x_2^2 / x_1), \tag{4.45}$$

the eigenvalue and the eigenfunction are obtained as

$$\lambda = \bar{a}\,(a_5^2 + a_6^2) / a_5 a_6,$$

$$\phi(x_1, x_2) = N.\,(x_1 x_2)\,.\exp\left[\frac{\bar{a}}{6a_5 a_6}\{(a_5^2 - 2a_6^2) x_1^2 + (a_6^2 - 2a_5^2) x_2^2\} + \frac{1}{2}(a_5\, x_1^2 x_2 + a_6\, x_1 x_2^2 + \bar{a}\, x_1 x_2)\right],$$

with the following three constraints,

$$a_{11} = 2\bar{a}^2\ (a_5^2 + a_6^2) / (a_5 a_6);$$

$$a_{20} = \bar{a}^2\,(4a_5^4 + 16a_6^4 - 7a_5^2\, a_6^2) / (36 a_5^2 a_6^2);$$

$$a_{02} = \bar{a}^2 (4a_6^4 + 16a_5^4 - 7a_5^2\ a_6^2) / (36a_5^2\ a_6^2)\,,$$

on the parameters. Interestingly, by taking either all the three or any one of the parameters $\bar{a}$, a_5, a_6 as negative, the potential (4.45) can offer a fairly well-defined bound state problem.

Note that in all the cases (*i*) – (*iv*) above the variables x_1 and x_2 with their odd powers only should be understood as $|x_1|$and $|x_2|$ with the corresponding odd powers mainly for the purposes of retaining the normalization of the eigenfunction. Further it can be noticed that the NC potentials of the type $V(x_1, x_2) = \sum_{i,j=0}^{n} b_{ij}\ x_1^i\ x_2^j$, $(i + j) \leq n$, and i, j are not zero simultaneously), in general, do not admit a normalizable solution of the equation unless some inverse harmonic and/or crossed terms in x_1 and x_2 are added to them.

Makarewicz [134] studied a sextic potential of the form

$$V(x_1,x_2) = \frac{1}{2}(\lambda_1 x_1^2 + \lambda_2\, x_2^2) + \eta\upsilon(x_1x_2)^2 + \frac{1}{2}\eta^2 (x_1x_2)^2\,(x_1^2 + x_2^2)\,, \tag{4.45a}$$

where λ_i, η and υ are the potential parameters and for $\eta<0$ this potential has two or four minima. For this case, one can explore a possibility of obtaining a closed-form solution of (4.30) in the form

$$\phi(x_1,x_2) = P(x_1,x_2).\exp\left[\frac{1}{2}(\omega_1 x_1^2 + \omega_2 x_2^2 + \eta\, x_1^2\, x_2^2)\right]. \tag{4.45b}$$

Here $P(x_1, x_2)$ is a polynomial in x_1 and x_2. It is argued that for certain constraints on the parameters λ_i, ω_i, υ and η, the bound state solutions for (4.45a) in the form (4.45b) do exist.

(*b*) *Exponential potentials:* If one departs slightly from the standard eigenfunction-ansatz method, then the exact solution of the SE for a class of NC exponential potentials can be obtained. In fact, instead of starting with a known form of the potential in advance, one can determine the potential itself for a specific form of $g(x_1, x_2)$ in (4.31) (with $f(x_1, x_2) = 1$) which is consistent with the solution of the SE. For example, for the form

$$g(x_1,x_2) = \beta_1 x_1 + \beta_2 x_2 + \beta_3.\exp(\alpha_1 x_1) + \beta_4.\exp(\alpha_2 x_2)$$
$$+ \beta_5.\exp(\alpha_3 x_1 + \alpha_4 x_2)\,, \tag{4.46}$$

eq. (4.30) yields [137] the expressions

$$v(x_1,x_2) = \beta_3^2\alpha_1^2.\exp(2\alpha_1x_1) + \beta_4^2\alpha_2^2.\exp(2\alpha_2x_2) +$$
$$\beta_3\alpha_1(2\beta_1+\alpha_1).\exp(\alpha_1x_1) + \beta_4\alpha_2(2\beta_2+\alpha_2).\exp(\alpha_2x_2)+$$
$$\beta_5^2\ (\alpha_3^2+\alpha_4^2).\exp(2(\alpha_3x_1+\alpha_4x_2)) + \beta_5(2\beta_1\alpha_3 + 2\beta_2\alpha_4 +$$
$$\alpha_3^2 + \alpha_4^2).\exp(\alpha_3x_1 + \alpha_4x_2) + 2\beta_3\beta_5\alpha_1\alpha_3.\exp[(\alpha_1+\alpha_3)x_1 +$$
$$\alpha_4x_2] + 2\ \beta_4\beta_5\alpha_2\alpha_4.\exp[\alpha_3x_1 + (\alpha_2+\alpha_4)x_2], \tag{4.47}$$

for the potential, and

$$\lambda = -(\beta_1^2 + \beta_2^2),$$

for the eigenvalues, respectively. Several special cases of the general form (4.47) could be of interest in various physical problems. For example, the cases corresponding to (*i*) $\alpha_3 = \alpha_4 = 0$ or $\beta_5 = 0$, (*ii*) $\alpha_1 = \alpha_2 = 0$ or $\beta_3 = \beta_4 = 0$, (*iii*) $\beta_1 = -\alpha_1/2$, $\beta_2 = -\alpha_2/2$; $\alpha_1 = \alpha_3$, $\alpha_2 = \alpha_4$ and (*iv*) $\beta_1 = \beta_2 = 0$, in (4.46) and (4.47) are analysed in Ref. [137]. Further restrictions on the arbitrary parameters α_i's and β_i's in (4.46) will give rise to the following classes of Morse-type NC but separable potentials in two dimensions:

$$v_1(x_1,x_2) = \alpha_1^2[\exp\{2\alpha_1(x_1-x_0)\} - 2.\exp\{\alpha_1(x_1-x_0)\}]$$
$$+ \alpha_2^2[\exp\{2\alpha_2(x_2-y_0)\} - 2.\exp\{\alpha_2(x_2-y_0)\}], \tag{4.48a}$$

with the minumum value $v_1(x_0,y_0) = -(\alpha_1^2 + \alpha_2^2)$ at the point (x_0, y_0), and

$$v_2(x_1,x_2) = (\alpha_3^2 + \alpha_4^2)\ [\exp\{2\alpha_3\ (x_1-x_0) + 2\alpha_4\ (x_2-y_0)\}$$
$$-2.\exp\{\alpha_3\ (x_1 - x_0) + \alpha_4\ (x_2 - y_o)\}], \tag{4.49a}$$

with the minimum value $v_2(x_0, y_0) = -(\alpha_3^2 + \alpha_4^2)$ at the point corresponding to the product variable $(XY) = \exp(\alpha_3x_1 + \alpha_4x_2)$ and characterized by $(X_0Y_0) = \exp(\alpha_3x_0 + \alpha_4y_0)$. The eigenvalues and the eigenfunctions corresponding to these potentials are found as

$$\lambda_1 = -(1/4)(\alpha_1^2 + \alpha_2^2), \tag{4.48b}$$

$$\phi_1 = N_1\ \exp[(1/2)\alpha_1x_1 + (1/2)\ \alpha_2x_2 - \exp\{\alpha_1(x_1-x_0)\} -$$
$$-\ \exp\{\alpha_2(x_2-y_0)\}] \tag{4.48c}$$

and

$$\lambda_2 = -(1/4)\ (\alpha_3^2 + \alpha_4^2), \tag{4.49b}$$

$$\phi_2 = N_2 \exp[(1/2)\,\alpha_3 x_1 + (1/2)\,\alpha_4\, x_2 - \exp\{\alpha_3 (x_1 - x_0)\} - \exp\{\alpha_4\,(x_3 - y_o)\}], \tag{4.49c}$$

respectively. Here the normalizations N_1 and N_2 can be obtained using (4.33). In analogy with the work of Markworth [138], the stability of two-dimensional monatomic or one-dimensional diatomic crystals can be studied using the above type Morse potentials. Interestingly, some of these special cases turn [137] out to be integrable by admitting second constant of motion besides the Hamiltonian.

(*c*) *Parabolic potentials*: Here we demonstrate as how the SE reduces to a standard-differential- equation whose solution already exists in the literature for a class of NC parabolic potentials following the work of Guha and Mukherjee [135]. In prolate spheroidal coordinates such a potential can be written as

$$V(\xi, \eta) = [U(\xi) + W(\eta)]/(\xi^2 - \eta^2),$$

for which the SE is separable. Note that for this system the angular momentum L^2 is not a constant of motion and an operator of the form,

$$\mathfrak{L} = L^2 + \Omega\,(\xi, \eta), \tag{4.50}$$

where $\Omega(\xi, \eta) = [(\xi^2 - 1)\, W(\eta) - (1 - \eta^2)\, U(\xi)]/(\xi^2 - \eta^2)$, is found to be the constant of motion. Clearly,

$$[H, \mathfrak{L}] = 0 \tag{4.51}$$

where $H = H_0 + V(\xi, \eta)$ with $H_0 = -\hbar^2\, \nabla^2/2m$. In parabolic coordinates, defined by $\xi = r - z$, $\eta = r + z$, $\phi = \varphi$ the operators ∇^2 and L^2 can be expressed as

$$\nabla^2 = \frac{4}{(\xi + \eta)}\left[\frac{\partial}{\partial \xi}\left(\xi\,\frac{\partial}{\partial \xi}\right) + \frac{\partial}{\partial \eta}\left(\eta\,\frac{\partial}{\partial \eta}\right)\right] + \frac{1}{\xi\eta}\,\frac{\partial^2}{\partial \varphi^2},$$

$$L^2 = \hbar^2\left[(\eta - \xi)\left(\frac{\partial}{\partial \xi} - \frac{\partial}{\partial \eta}\right) + \eta\xi\left(\frac{\partial}{\partial \xi} - \frac{\partial}{\partial \eta}\right)^2 + \frac{(\eta + \xi)^2}{4\eta\xi}\,\frac{\partial^2}{\partial \varphi^2}\right],$$

and subsequently the condition (4.51) yields the form of the potential as

$$V(\xi, \eta) = -2\,(C_2\xi + C_1\eta)/\{\xi\eta(\xi + \eta)\}. \tag{4.52}$$

Guha and Mukherjee [135], however, investigate the bound state problem in a potential

$$\tilde{V}(\xi, \eta) = -[2a/(\xi + \eta)] + 2\,(C_2\xi + C_1\eta)/\{\xi\eta(\xi + \eta)\}, \tag{4.52$'$}$$

which also involves a Coulomb part. For $E < 0$, the solution of the SE can be written as

$$\psi(\xi, \eta, \varphi) = N f_1(\xi) f_2(\eta)\, e^{\pm im\varphi},$$

where f_i ,($i = 1, 2$) satisfy

$$\frac{1}{f_i}\frac{\partial}{\partial x_i}\left(x_i \frac{\partial f_i}{\partial x_i}\right) - \frac{1}{4}\left\{4\mu\frac{C_i}{\hbar^2} + m^2\right\}\frac{1}{x_i} + \frac{\mu E}{2\hbar^2}x_i = -\sigma_i, \qquad (4.53)$$

where $x_1 = \xi$, $x_2 = \eta$ and the constants σ_i satisfy the condition $\sigma_1 + \sigma_2 = \mu a/\hbar^2$. Finally, the solution of (4.53) is obtained as [135]

$$f_i = \zeta_i^{|S_i|/2}.\exp\left(-\frac{1}{2}\zeta_i\right).L_{\upsilon_i}^{|S_i|}(\zeta_i),$$

provided $v_i = n_i = \lambda_i - \frac{1}{2}(|S_i| + 1)$ are non-negative integers. Here $L_{\upsilon_i}^{|S_i|}$ are the Laguerre functions and $\lambda_i = \sigma_1/\alpha$ with $\alpha = (-2\mu E/\hbar^2)^{1/2}$ and $\zeta_i = ax_i$. The corresponding eigenvalues can be obtained from the expression

$$E = -\frac{\mu a^2}{2\hbar^2}\left[n_1 + n_2 + 1 + \frac{1}{2}\sqrt{m^2 + (4\mu C_1)/\hbar^2} + \frac{1}{2}\sqrt{m^2 + (4\mu C_2)/\hbar^2}\right]^{-2}.$$

immediate applications of these results mainly to the cases of Hartmann and Aharonov-Bohm potentials are worth noting.

4.5 APPROXIMATION METHODS

In the absence of availability of the exact solution of a problem in quantum mechanics three main approximation methods are generally employed for this purpose, namely perturbation, variational and WKB methods. Further, approximation methods for the 2D and 3D problems are not as frequently developed and used as for the 1D problems inspite of the fact that the closed-form solutions (discussed in the previous sections) have become available to check the accuracy and efficiency of these methods.

Depending upon the type of the problem any one or all of the three approximation methods mentioned herein above have been used for the 1D problems. Besides that the applications of the WKB method are well-known to the 1D problems (or to the cases where the higher dimensional SE is separable into total differential equations each involving only a single independent variable), the same has also been extended to a restricted class of higher dimensional systems in the form of what is known Einstein-

Brillouin-Keller (EBK) quantization rule [141]. In what follows, we make a few remarks about these methods with reference to their applicability to two or higher dimensional systems.

In fact, the variational method for a 2D problem would require an evaluation of the minimum of energy from the expression

$$E = \int_{-\infty}^{\infty}\int_{-\infty}^{\infty} \psi^*(x_1, x_2)\,(H(-i\hbar\partial/\partial x_1, -i\hbar\partial/\partial x_2, x_1, x_2)\,\psi(x_1, x_2)dx_1dx_2\,/$$

$$(\int_{-\infty}^{\infty}\int_{-\infty}^{\infty} |\psi((x_1, x_2)|^2 dx_1dx_2). \tag{4.54}$$

Here, for the purposes of choosing the form of the trial wavefunction, $\psi(x_1,x_2)$, the closed form solutions discussed in the previous sections, no doubt, may turn out to be instructive but such a choice, depending again on the type of the 2D problem, is limited in view of the constraints on the potential parameters. Ideally speaking, using the variational method a 2D problem can be solved if the trial wavefunction involves only one parameter and the same is normalizable. This, however, is generally not the case with NC problems. Very often more than one parameter are bound to appear in the trial wavefunction for the NC problem and consequently the complexity in evaluating the minimum E from (4.54) increases even if the trial wavefunction is normalizable. Thus, for the NC 2D or 3D problems the most suited approximation method is the perturbation one.

The EBK quantization in the context of higher dimensional systems can be described better in terms of action-angle variables. In this case, one considers a completely integrable TID Hamiltonian system in N dimensions and a wavefunction of the form $\psi = A \exp(iS/\hbar)$, where the action function S is given by [141]

$$S(\vec{q}, \vec{J}) = \int_{\vec{q}_0}^{\vec{q}} \vec{p}\,(\vec{q}', \vec{J})\, d\vec{q}'. \tag{4.54a}$$

Here $\vec{q}_0$ is some (arbitrary) initial point; the action variable $\vec{J}$ and the position variable $\vec{q}$ are respectively constituted from $(J_1, J_2, \ldots J_N)$ and $(q_1, q_2, \ldots, q_N)$ which are defined along each dimension. The important point to be noted here is that the action function S in this case becomes multivalued function of $\vec{q}$, since the conjugate momentum $\vec{p}$ itself is now multivalued (unlike the 1D case where $p(q,J) = \pm\,[2m\,(H(J) - V(q))]^{1/2}$, is two-valued function of q). Thus, the wavefunction of the form $\psi = A.\exp(iS/\hbar)$ must be summed over all possible branches of S leading to

$$\psi(\vec{q}) = \sum_r \det\left|\frac{\partial^2 S_r}{\partial q_j \, \partial J_k}\right|^{1/2} . \exp(iS_r(\vec{q}, \vec{J}) / \hbar), \qquad (4.54b)$$

where the sum over r denotes the sum over branches (unlike the WKB case where just two-branch sum is associated with the 1D bounded motion). The condition of single-valuedness on the wavefunction (4.54b) finally takes [141] the form

$$\frac{1}{\hbar}\oint_{C_k} \vec{p}\,(\vec{q}', \vec{J})\, d\vec{q}' - \alpha_k \frac{\pi}{2} = 2n_k \pi,$$

where for an N-dimensional torus there are N-topologically distinct classical paths C_k, and the numbers α_k's are referred to as Maslov indices in the literature. This will lead to the generalized multidimensional quantization condition

$$J_k = \oint_{C_k} \vec{p}.d\vec{p}' = 2\pi\hbar[n_k + \frac{1}{4}\alpha_k],$$

which is usually known as EBK quantization rule.

The use of approximation techniques to study the power potentials (particularly the even power potentials) in 1D and 2D has been an active area of interest for a long time now (see, for example, the review by Hioe et al. [139] and Ref. [98]) mainly because such potentials play [140] an important role in various physical problems besides their immediate implications in quantum field theory. Recently, Witwit in a series of papers [142a, 142b] has used both perturbative and nonperturbative techniques to study the NC even power potentials in 2D and 3D involving quartic, sextic and octic anharmonicities. An improved Hill-determinant method in terms of vector recurrence relations has also been used [143] recently to study polynomial and nonpolynomial potentials in 1D. Eigenvalues and eigenfunctions are obtained by Witwit to a fair degree of accuracy and for a wide range of values of the perturbation parameter. In particular, the potentials investigated by him are of the form

$$V(x_1, x_2) = \frac{1}{2}(x_1^2 + x_2^2) + \frac{1}{2}\lambda(b_{60}x_1^6 + b_{06}x_2^6 + 3b_{42}x_1^4x_2^2 + 3b_{24}x_1^2x_2^4),$$

$$V(x_1, x_2) = \frac{1}{2}(x_1^2 + x_2^2) + \frac{1}{2}\lambda(b_{80}x_1^8 + b_{08}x_2^8 + 6b_{44}x_1^4x_2^4 +$$

$$+ 4b_{62}x_1^6x_2^2 + 4b_{26}\, x_1^2x_2^6), \qquad (4.55)$$

in 2D[142a] and

$$V(x_1,x_2,x_3)=\frac{1}{2}(x_1^2+x_2^2+x_3^2)+\frac{1}{2}\lambda(b_{400}x_1^4+b_{040}x_2^4+b_{004}x_3^4$$

$$+\ 2b_{220}x_1^2x_2^2\ +\ 2b_{202}x_1^2x_3^2\ +\ 2b_{022}x_2^2x_3^2),$$

$$V(x_1,x_2,x_3)=\frac{1}{2}(x_1^2+x_2^2+x_3^2)+\frac{1}{2}\lambda(b_{600}x_1^6+b_{060}x_2^6+b_{006}x_3^6+$$

$$+\ 6b_{222}x_1^2x_2^2x_3^2\ +\ 3b_{402}x_1^4x_3^2\ +\ 3b_{420}x_1^4x_2^2\ +\ 3b_{042}x_2^4x_3^2\ +$$

$$+\ 3b_{240}x_1^2x_2^4\ +\ 3b_{204}x_1^2x_3^4\ +\ 3b_{024}x_2^2x_3^4), \tag{4.56}$$

in 3D[142b]. Here λ is the perturbation parameter. Note that in all the above cases, while the perturbation is essentially the NC part, the unperturbed part is the central one. For further details the reader is referred to the work of Witwit.

4.6 EXACT SOLUTIONS AND THE ROLE OF CONSTRAINTS ON POTENTIAL PARAMETERS

Exact solutions of the SE discussed in this Chapter concern mainly a few eigenvalues of the system and very often that too with certain constraints on the potential parameters. If the exact solution of the SE for a given system is obtained without any constraint (like that for the Morse central potential (4.28), then it is an ideal situation; otherwise the role of these constraints is very crucial not only in deciding the ground and higher excited states but even in obtaining the bound states for the system. In order to demonstrate this fact we consider the example of the central potential (4.25). As far as the ground state in this potential is concerned the same can be obtained [97] in an unambiguous manner by using the constraint (4.26′). For the excited states, however, one makes [115, 144] an ansatz for the wavefunction as

$$\phi(r)=(1+\alpha_1 r^2+\alpha_2 r^{-2}).\exp\left(\frac{1}{2}\alpha r^2+\frac{1}{2}\beta r^{-2}+\delta_1\,\ell n\ r\right). \tag{4.57}$$

Using this form in (4.3) and subsequently rationalizing the resultant expression one obtains [115]

$$\beta=-\sqrt{c}\ ;\ \alpha=-\sqrt{a}\ , \tag{4.58a,b}$$

$$E_1 = \sqrt{a}\,(5 + 2\delta_1)\ ;\ \delta_1 = (b + 7\sqrt{c})/2\sqrt{c}, \qquad (4.58c, d)$$

$$\alpha_1 = 4\sqrt{a}\,/\,(\eta_\ell - 4\delta_1 - 2)\ ;\ \alpha_2 = 4\sqrt{c}\,/\,(\eta_\ell + 4\delta_1 - 6), \quad (4.58e,f)$$

with the constraint

$$\eta_\ell[(\eta_\ell - 4)^2 - 4(2\delta_1 - 1)^2] + 64\sqrt{ac}(\eta_\ell - 4) = 0, \qquad (4.59)$$

where

$$\eta_\ell = \ell(\ell + 1) + 2\alpha\beta - \delta_1^2 + \delta_1.$$

From a detailed analysis of the problem Varshni [117] has argued that there are four sets of solutions for δ_1, α_1,α_2 and E which satisfy eqs. (4.58a) – (4.58f) and accordingly this leads to different forms of the constraint (4.59) even for a = 1 and ℓ = 0. As a result a proper ordering of energy of the excited states, which otherwise was not possible [115], has been obtained.

To highlight further the role of constraints we mention the case of another potential, $V(\mathrm{x}) = Ax^6 + bx^2$, in 1D which has recently been studied by Bender and Turbiner [145] and also by de Souza Dutra et al. [146]. Starting from particular analytic solutions (which are valid when certain relations between the parameters A and B of this potential hold) de Souza Dutra et al. [146] have introduced an approximate expression for the eigenenergies. This expression, though capable of providing the complete spectrum, clearly indicates the deviation from the constraining relation in terms of the numerical accuracy.

It may be mentioned that such analyses vis-a-vis energy ordering of the excited states become relatively more difficult for the NC potentials even in 2D. This is mainly because the existence of the absolute minimum (in the sense that it should conform to all independent variables) itself is a problem with most of the NC potentials. For an analysis of the constraints with reference to the existence of the bound states in an NC potential in 2D, we refer to the work of Makerewicz [134].

4.7 ROLE OF DYNAMICAL SYMMETRIES IN QUANTUM MECHANICS

With a view to obtaining exact solution of the SE one can also explore dynamical symmetries in nonrelativistic quantum mechanics. In particular, one can ask what potential functions in general possess a symmetry group which is not of explicit geometric origin (i.e. not directly related with the symmetry of physical space and time). Such studies, carried out by several authors [147,168], have provided several interesting features of NC (but very often separable) potentials in both 2D and 3D.

In this case one looks for a Hamiltonian $H = -(\hbar^2/2m)\ \nabla^2 + V(x_1,x_2)$, where $\nabla^2 = (\partial^2/\partial x_1^2 + \partial^2/\partial x_2^2)$, for which there exists a set of linear differential operators L_i commuting with H, namely $[L_i,H] = 0$. At the same time one also finds the constants of motion for the corresponding system. Winternitz et al [147] consider L_i's of order not higher than two in momenta for 2D systems and Makarov et al. [147] considered the same for 2D and 3D systems. The constants of motion of third order are also investigated by Datta Majumdar and Englefield [168] within this framework. For example, for the first order (in momenta) rather trivial (in the sense that the system possesses geometric symmetry) case, the constant of motion L' turns out to be

$$L' = a(x_2p_1 - x_1p_2) + cp_1 - bp_2\,, \tag{4.60}$$

corresponding to the potential, $V(x_1,x_2) = V(X)$, where $X = \frac{1}{2}\, a\,(x_1^2 + x_2^2) + bx_1 + cx_2$, and $V(X)$ is an arbitrary function of X. On the other hand, for the quadratic form of the constant of motion,

$$L = aM^2 + b(p_1M + Mp_1) + c(p_2M + Mp_2) + dp_1^2 + ep_2^2 + f\,p_1p_2 + \varphi(x_1, x_2), \tag{4.61}$$

where $M = x_1p_2 - x_2p_1$ and $\varphi(x_1, x_2)$ is an arbitrary function, Winternitz et al [147], in general, have shown that such a constant of motion can exist only if there exists a coordinate system (q_1, q_2) in which the variables separate.

For the following four special forms of L, the corresponding potentials derived are

(i) $L = p_1^2 - p_2^2 - 2f(x_1) + 2g(x_2)$; $V = f(x_1) + g(x_2)$,

(ii) $L = M^2 + \varphi(x_1,x_2)$; $V = [f(a) + g(\varphi)]/(\exp(2a))$,

with spherical coordinate system defined by $x_1 = e^a\cos\varphi$, $x_2 = e^a\sin\varphi$

(iii) $L = p_1M + Mp_1 + \varphi(x_1,x_2)$; $V = [f(\xi_1) + g(\xi_2)]/(\xi_1^2 + \xi_2^2)$.

with parabolic coordinate system defined by $x_1 = \frac{1}{2}(\xi_1^2 - \xi_2^2)$, $x_2 = \xi_1\xi_2$, and

$$\varphi(x_1,x_2) = [f(\xi_1)\xi_2^2 - g(\xi_2)\,\xi_1^2]/(\xi_1^2 + \xi_2^2).$$

(iv) $L = M^2 + \frac{1}{2}\ell^2(p_1^2 - p_2^2) + \varphi(x_1,x_2)$; $V = [f(\sigma) + g\,(\rho)]/$

$(\cos^2\sigma - \cos h^2\rho)$

with elliptic coordinate system, $x_1 = \ell \cosh\rho \cos\sigma$, $x_2 = \ell \sinh\rho \sin\sigma$, and $\varphi(x_1, x_2) = -\ell^2[(\cosh^2\rho + \sinh^2\rho) f(\sigma) + (\cos^2\sigma - \sin^2\sigma)g(\rho)]/(\cos^2\sigma - \cos^2\rho)$.

From the point of view of applying the theory of Lie groups, the systems which (*i*) possess two or more independent operators of the type

$$L = \varphi_1 p_1^2 + \varphi_2 p_2^2 + \varphi_3 p_1 p_2 + \varphi_4 p_1 + \varphi_5 p_2 + \varphi_6, \tag{4.62}$$

where φ_i's are arbitrary functions of x_1 and x_2, and (*ii*) admit separation of variables in two coordinate systems (either different or of the same type, but displaced or rotated w.r.t. each other) are of special interest. A number of such systems in 2D and 3D are found [147]. Naturally, not only for these systems but also for the systems (*i*) – (*iv*) mentioned above, closed-form solution of the SE alongwith complete spectrum of the eigenvalues is obtained. For details we refer to these works.

In analogy with (4.62), Datta Majumdar and Englefield [168] consider a third order operator

$$L = \varphi_1 p_1^3 + \varphi_2 p_1^2 p_2 + \varphi_3 p_1 p_2^2 + \varphi_4 p_2^3 + \varphi_5 p_1^2 + \varphi_6 p_1 p_2 + \varphi_7 p_2^2 + \varphi_8 p_1 + \varphi_9\, p_2 + \varphi, \tag{4.63}$$

for the potential $V(x_1, x_2)$. Using the condition $[L, H] = 0$, the form of the operator L is found to be

$$L = M^3 + (1\text{-}3G)\, M - 3/2\, G', \tag{4.64}$$

corresponding to the form of the potential $V(r, \theta) = F(r) + r^{-2}\, G(\theta)$, which allows the separation of the SE in polar coordinates. Here, $M = \partial/\partial\theta$ and $G' = dG/d\theta$.

4.8 CONCLUDING DISCUSSION

Under the title of quantum mechanics of NC TID systems in this chapter, our aim has been to look for analytic solutions of the SE and to enlarge the catalogue of exactly solvable NC potentials in 2D. For this purpose we are often guided by the methods available for the 1D systems. No doubt several other quantum mechanical aspects could have been included in this survey but thinking that once the eigenvalues and eigenfunctions of a system become available then the study of all other aspects of the system becomes secondary; the study of other aspects does not find a place to a desired extent in this survey. Moreover, on every front of study in quantum mechanics, the NC potentials are going to pose difficulty of one type or the other. Some of them are highlighted in this survey (cf. Sect. 4.3).

It may be recalled that if a system involves NC forces then the angular momentum of such a system is not a constant of motion. In such a situation the quantum eigenstates need to be constructed corresponding to the second (besides the Hamiltonian) constant of motion for the 2D system, provided such a constant of motion exists and becomes available. Again, if this second constant of motion of a 2D system does not exist (i.e. if the system is not integrable) several conceptual problems arise at the foundation level of quantum mechanics of these systems. In fact, these general remarks made here are also valid for the study of quantum aspects of TD systems in 2D which will be pursued in the next chapter. However for such further details the reader is referred to the review by Eckhardt[10].

A few words about the methods used for obtaining analytic solutions of the SE are worth mentioning. In general, the available methods for 2D systems are not found as viable as for the 1D systems in the sense that they are more case-oriented. In fact, all the methods used for the 1D systems are not yet extended to the study of 2D systems. (It does not appear to be an easy task either.) For example, the supersymmetric factorization method need to be extended to the 2D systems. Also, an extension of the quasidynamical Lie algebraic method (cf. Sect. 4.2.2) has not yet been carried out to 2D systems at the quantum level. On the other hand, at the classical level such an extension has offered [85] several new features for the 2D and 3D systems. Even in the 1D case this method has not suggested more than one form (cf. eq. (4.9)) of the exactly solvable potential. Following the work of Burrows et al [124] it appears that the choice of an appropriate realization of generators of the algebras (either of $so(3)$ or of $so(2,1)$ is of great significance as far as the bound-state solutions of the SE are concerned. Further, the functional dependence of the Hamiltonian on these generators is responsible for giving rise to a full or a part of the bound-state spectrum. However, such studies for the 2D systems have not yet been pursued.

It may be noted that the approximation methods (both perturbative and nonperturbative) have been used successfully (in the sense of numerical accuracy) to study the 1D problems during the last few decades. However, not many attempts have been made so far to study the 2D systems using these methods. In fact, such methods are yet to be developed for the 2D systems in general.

What we have obtained in this chapter are the analytic solutions of the SE which only correspond to a few bound states and that too with certain constraints on the potential parameters. From the point of view of applications even these solutions, no doubt, are useful (cf. Sect. 4.1), but still the knowledge of complete eigenvalue spectrum, may be at the cost of eliminating [146] the constraining relations, is desirable for a complete understanding of the system. Although such attempts have been made [146] for the 1D systems, studies such as these for the 2D systems, however, are going to be a tedius task. Further, in this context the nature of constraining relations among the parameters of the potential will be very important.

5

Quantum Mechanics of Noncentral Time Dependent Systems

5.1 INTRODUCTION

In the previous Chapter we have seen that there are not many potentials for which the stationary state SE admits an exact solution. In fact, the number of the systems involving explicit time-dependence, for which the corresponding TDSE admits an exact solution is further less. Note that the definition of the exact solution in this Chapter will be more refined than that in the previous Chapter (cf. Sect. 4.2.2) in the sense that we shall look for the complete spectrum of the eigenvalues. Further, this scarcity of exactly solvable systems often forces one to use rather involved approximation methods to solve the TD problem in this case also. However, we shall avoid here the discussion of these methods which are easily available in several textbooks (see, for example, Refs. [8] and [99]) and concentrate mostly on the methods used to obtain an exact solution of the problem. As a matter of fact in order to obtain an exact solution of the TDSE for a TD potential in 1D, the role of a TD invariant, if it exists, cannot be ignored in this context. (This is mainly because the Hamiltonian in this case is not the constant of motion.) It may be reminded (cf. Chap. 1) that the number of such independent invariants (provided they exist) will increase if one considers the TD systems in higher dimensions from the point of view of their integrability in the classical sense.

In this Chapter, mainly two aspects pertaining to exact solutions of some TD problems will be discussed: In one case the wavefunction will be obtained in Schrodinger representation using the Feynman propagator approach of Khandekar and Lawande [19] and in the second, we make use of a modified version of the eigenfunction-ansatz method of Chap. 4 to obtain the solution of the TDSE in 1D and 2D. A brief survey of the time-evolution operator and Lie algebraic approach [148-153] is also made in Sect. 5.2. This rather fundamental approach has been used so far to study 1D systems only. The arrangement of the present chapter is as follows: In the next section, we make a survey of quantum mechanics of TD systems in 1D. In Sect. 5.3, the Feynman propagator approach, mostly applied so far to 1D problems by Khandekar and Lawande in a series of papers [19, 175, 176], is briefly discussed. TD systems in 2D are investigated in Sect. 5.4. Finally, the concluding remarks are made in Sect. 5.5.

5.2 SURVEY OF ONE-DIMENSIONAL TD SYSTEMS

5.2.1 General Survey

As mentioned before the exact solutions of SE, with the Hamiltonians involving explicit time-dependence, are available only in a few cases. In the past there had been considerable interest [64, 80, 81, 147-160, 200] in the search not only of such exactly solvable systems but also of the techniques to obtain them. Further, in this context, not only the systems in 1D are often investigated but even within this the case of harmonic oscillator with variable frequency (a simplest possible system which has a lot of physical relevance) has been the test case for various methods developed so far. As far as the merit or the viability of these techniques is concerned, it may be mentioned that the reproduction of the results for HO is a necessity but not sufficiency.

About four decades ago, Husimi [200] studied the quantal treatment of the forced TDHO in 1D with reference to a variety of phenomena. While some of his results pertain to the exact treatment of the explicit time-dependence of the system, many of them are obtained for the adiabatic or perturbative case. In fact, some of his results, related to the subharmonic resonances, are now well taken care of through the coherent state formalism [196] and also by introducing [198] a linear (but complex) invariant in the formalism. For details we refer the reader to Sect. 7.2.6 (d).

Lewis and Riesenfeld [64] (LR) developed a general theory of explicitly TD invariants for quantum systems. They obtained a simple relation between eigenstates of the necessary invariant and solutions of the corresponding SE. According to LR, for a quantum system characterized by a TD Hamiltonian operator $H(t)$ and a Hermitian invariant operator $I(t)$ the general solution of the TDSE

$$i\hbar\,(\partial\Psi(x,t)/\partial t) = H(t)\,\Psi(x,t), \tag{5.1}$$

is given by

$$\Psi(x,t) = \sum_n c_n \exp(i\alpha_n(t)).\psi_n(x,t). \tag{5.2}$$

Here $\psi_n(x,t)$ are the normalized eigenfunctions of the operator $I(t)$ and

$$I\,\psi_n(x,t) = \lambda_n\,\psi_n(x,t), \tag{5.3}$$

where the eigenvalues λ_n are time independent. An important restriction on the form of $I(t)$ here is that it does not involve the time derivatives. Note that the expansion coefficients c_n are constants while the TD phases $\alpha_n(t)$ are obtained from

$$\hbar(\partial\alpha_n(t)/\partial t) = \langle\psi_n\,|\,i\hbar\,\partial/\partial t - H\,|\,\psi_n\rangle. \tag{5.4}$$

LR solved explicitly for λ_n, $\alpha_n(t)$ for the TD quantum HO described by the Hamiltonian (cf. eq. (3.4)) operator,

$$H = \frac{1}{2}\,[p^2 + \omega^2(t)\,x^2], \tag{5.5}$$

where p and x are now q-numbers. It is worth recalling the forms of $I(t)$ and of the auxiliary equation corresponding to (5.5) (cf. eqs. (3.5) and (3.6))

$$I(t) = \frac{1}{2}\,[k\,(x/\rho)^2 + (\dot{x}\rho - x\dot{\rho})^2], \tag{5.6}$$

$$\ddot{\rho} + \omega^2(t) = k\rho^{-3}. \tag{5.7}$$

Hartley and Ray [154] make use of a unitary transformation

$$\psi_n'(x,t) = \mathcal{U}\,\psi_n(x,t), \tag{5.8}$$

with $\qquad \mathcal{U} = \exp\,(-i\dot{\rho}x^2/(2\,\hbar\rho))\,,$

to reduce the solution of (5.1) to the solution of a TID one-dimensional SE of the form

$$I'\,\phi_n(\sigma) = \lambda_n\,\phi_n(\sigma), \tag{5.9}$$

where $I' = \mathcal{U} I\,\mathcal{U}_+$; $\psi_n'(x,t) = \rho^{-1/2}\,\phi_n(\sigma)$, and $\sigma = (x/\rho)$. Here the $\rho^{-1/2}$ factor is introduced mainly to retain the normalization of $\phi_n(\sigma)$ as

$$\int \psi_n'^{\,*}(x,t)\,\psi_n'(x,t)\,dx = \int \phi_n^*(\sigma)\,\phi_n(\sigma)\,d\sigma = 1.$$

In this new (primed) representation, the phases $\alpha_n(t)$ are determined [155] from

$$\alpha_n(t) = -(\lambda_n / \hbar) \int^t dt/\rho^2 , \qquad (5.10)$$

where the function ρ is given by (5.7). From the above analysis Hartley and Ray arrived at what they call as a quantum-mechanical superposition law defined by $x = \rho u$, where x is the general solution of the equation of motion corresponding to (5.5) and u is the general solution of the differential equation

$$\frac{d^2u}{d\tau^2} + k\, u = 0 ,$$

with $d\tau = dt/\rho^2$. Later, Hartley and Ray [155] also applied these results to obtain the quantum-mechanical superposition law for more general TD systems.

Using the LR theory within the framework of Feynman propagator approach, Khandekar and Lawande [156] studied the quantum problem related to the TD Hamiltonian

$$H = \frac{1}{2}\, p^2 + \frac{1}{2}\omega^2\,(\mathrm{t})\; x^2 + b/x^2 . \qquad (5.11)$$

We shall return to some of these discussions in the next Section. While Leach [157] obtained the generalized invariants for the quadratic Hamiltonians, Camiz et al [158] using the Schrodinger formalism studied a system similar to (5.11).

5.2.2 Time-evolution Operator and Lie Algebraic Approach

This approach which is essentially based on the work of Wichmann [148] and of Wei and Norman [149] is carried out mostly for quadratic TD Hamiltonians. It has recently been cast into an applicable form by Wolf and Korsch [150] and by Fernandez [151]. In a nut-shell, in this approach, one looks for the solution of TDSE (5.1) by introducing a time-evolution operator $U(t)$ which satisfies the SE ($\hbar$ = 1)

$$i\;(dU\,(t)/dt) = H(t)\;U(t),\; U\,(0) = 1. \qquad (5.12)$$

Further, $H(t)$ is written as a linear combination of TID operators Γ_j with TD scalar-valued functions $h_j(t)$ as

$$H(t) = \sum_{j=1}^{M} h_j(t)\,\Gamma_j . \qquad (5.13)$$

When the operators are required to generate a finite-dimensional Lie algebra L with basis $\{\Gamma_1, \Gamma_2, \ldots \Gamma_N\}$, $N \geq M$, then the solution of (5.12) can be expressed in terms of this algebra. According to Wei and Norman [149], the splitting of Lie algebra L into smaller algebras reduces the task of finding $U(t)$. This makes the Lie algebraic structure theory more viable in the sense that many problems differing at the first sight may be shown to have isomorphic algebras and therefore have structurally identical solutions. For this purpose two convenient choices of $U(t)$ are mentioned by Wolf and Korsch [150], namely

$$U(t) = \exp\left[\sum_{j=1}^{N} f_j(t)\,\Gamma_j\right] \text{ and } U(t) = \prod_{j=1}^{N} \exp\,[g_j(t)\,\Gamma_j], \qquad (5.14)$$

where both f_j's and g_j's are determined by nonlinear equations in general. Note that in this approach the operators Γ_j's are essentially expressed in terms of creation and annihilation operators. Using the second form of $U(t)$ while Wolf and Korsch [150] have studied the problem of coupled TD oscillators, a generalized TD quadratic Hamiltonian has also been investigated by them and also by Fernandez [151].

This Lie algebraic approach offers an efficient and systematic tool to study several physical properties of a TD system. For example, matrix elements and transition probabilities can easily be written in terms of the solutions of quantum-mechanical equations of motion and they turn out [150, 152] to be independent of the form chosen for $U(t)$.

Now the question arises as to what is the connection between LR approach discussed in Sect. 5.2.1 and Wei and Norman approach discussed above. Recently, Salmistraro and Rosso [153] have tried to find an answer to this question in terms of the invariant formalism. The latter is found to provide a link between the two approaches. Further, their discussions are based on a general representation of invariants which is valid for an arbitrary Hamiltonian and directly involves the time-evolution operator $U(t)$. Following their work, a simple representation theorem is worth presenting here.

Theorem: Every invariant operator $I(t)$ may be expressed in the form

$$I(t) = U(t)\, I(0)\, U^{-1}(t), \qquad (5.15)$$

where $U^{-1}(t)$, defined in an interval $0 \leq t \leq T$, is the inverse of $U(t)$ which satisfies (5.12).

Proof: Note that the Hamiltonian $H(t)$ is a continuous function of t and the invariant operator $I(t)$, by definition, is given by

$$\partial I(t)/\partial t \; - \; i\,[I(t), \;\; H(t)] = 0. \qquad (5.16)$$

Further an equation for $U^{-1}(t)$ can be obtained from (5.12) as

$$i\,(\partial U^{-1}(t)/\partial t) = -\,U^{-1}(t)\;\; H(t). \qquad (5.17)$$

Now, on operating (5.16) by U^{-1} from left and by U from right, one obtains the expression

$$U^{-1} \frac{\partial I}{\partial t} U - i\, U^{-1}\, IHU + i\, U^{-1}\, HIU = 0\,,$$

which after using (5.12) and (5.17), immediately yields [153]

$$\partial(U^{-1}\, IU)/\partial t = 0,$$

or

$$U^{-1}\, IU = \mathcal{C}$$

where $\mathcal{C}$ given by $\mathcal{C} = I\,(0)$ since $U\,(0) = 1$, is a TID operator. This obviously implies the result (5.15). Conversely, it is not difficult to verify that if $U\,(t)$ solves eq. (5.12), then an operator of the form (5.15) satisfies eq. (5.16).

5.2.3 Exact Solution to TDSE in 1D

In the past, several authors [80, 81, 154-160] have considered exact solution to the TDSE ($\hbar = \mu = 1$)

$$[\frac{1}{2}\,(\partial^2/\partial x^2) + V(x,\, t)]\;\Psi(x,\, t) = i\;(\partial\Psi(x,\, t)/\partial t), \tag{5.18}$$

in 1D using different techniques. For example, Khandekar and Lawande [159] and Hartley and Ray [154, 155] employ the method of TD invariants as suggested by LR [64]. Burgan et al. [81] make use of the symmetry group of the SE under the rescaling of the space and time variables along with the unitary transformation of the wavefunction. Use of the algebra of symmetries of the TDSE (5.18) is made by Truax [160] to study the potential

$$V\,(x,\, t) = g_1\,(t)\; x^2 + g_1\,(t)\; x + g_o\,(t). \tag{5.19}$$

Ray [80] has considered the generalization of the method of Burgan et al [81] to solve the problem of this TD quadratic potential. While we have already outlined the work of Hartley and Ray [154] in Sect. 5.2.1 and will discuss some further generalizations of the methods of Khandekar and Lawande and of Burgan et al, in the following Sections, here following Ray [80] we briefly discuss the solution of (5.18) for the potential (5.19).

The method of Ray [80] is essentially carried out in two stages. In the first stage one performs a scale and phase transformation of the dependent variable and a scale transformation of the independent space-time variables as follows:

$$\Psi\;(x,\; t)\; = B(t).\; \exp\,(i\phi\;(x,\; t))\;.\;\psi\,(x,t), \tag{5.20}$$

$$x' \; = x/C(t) + A(t);\; t' \; = D(t), \tag{5.21a,b}$$

where $B(t)$ is a TD normalization and $A(t)$, $C(t)$ and $D(t)$ are arbitrary functions of t. Under these transformations eq. (5.18) becomes

$$-\frac{1}{2}[iB\phi_{xx}\psi - B(\phi_x)^2\psi + (2iB/C)\ \phi_x\ \psi_{x'} + (B/C^2)\ \psi_{x'x'}] + VB\psi$$

$$= i\dot{B}\psi - B\phi_t\psi + iB\ [-(x\dot{C}/C^2) + \dot{A}]\ \psi_{x'} + iB\dot{D}\ \psi_{t'} \qquad (5.22)$$

where (x,x') and (t,t') subscripts indicate partial differentiation w.r.t. these variables and the dot denotes t differentiation. The arbitrary functions in the transformations (5.20) and (5.21) can be fixed by setting some of the additional terms in (5.22) equal to zero and subsequently by demanding the form of the TDSE (5.18) to be invariant under this transformation. In terms of the functions A and C, this will yield [80]

$$\phi = \frac{1}{2}\ (x^2\dot{C}\ /\ C) - \dot{A}\ C\ x, \qquad (5.23)$$

$$t' \equiv D\ (t) = \int dt/C^2\ ,\ B = 1/C^{1/2},$$

and a modified form $V'\ (x,t)$ of the potential as

$$V' = C^2V + \frac{1}{2}\ C\ \ddot{C}\ x^2 - (\ddot{A}\ C^3 + 2\dot{A}\dot{C}\,C^2)\,x + \frac{1}{2}C^4\ \dot{A}^2. \qquad (5.24)$$

Now for the form (5.19) of $V(x, t)$, the modified potential (5.24), in terms of the primed variables takes the form

$$V' = (g_2C + \frac{1}{2}\ \ddot{C})C^3\ x'^2 - [\ddot{A}C^4 + 2\ \dot{A}\dot{C}C^3 - g_1C^3 +$$

$$2\ (g_2C + \frac{1}{2}\ \ddot{C})\ AC^3]\,x' + F(t'), \qquad (5.25)$$

where

$$F(t') = (g_2C + \frac{1}{2}\ddot{C})\ A^2C^3 + 2\,AC\,(\ddot{A}\,C^3 + 2\,\dot{A}\dot{C}C^2 - g_1C^2) + \frac{1}{2}C^4\ \dot{A}^2 + g_0\ C^2.$$

We choose the function C to satisfy the equation

$$\ddot{C} + 2g_2\,(t)\ C = k/C^3, \qquad (5.26)$$

where k is an arbitrary constant. The function A is so chosen that the term linear in x' in (5.25) vanishes and thus giving rise to

$$\ddot{A} + 2(\dot{A}\dot{C}/C) + k(A/C^4) - g_1/C = 0. \tag{5.27}$$

Note that eq. (5.25) now takes the simple form

$$V' = \frac{1}{2} kx'^2 + F(t'), \tag{5.28}$$

with $F(t') = -\frac{1}{2} kA^2 + \frac{1}{2} C^4 \dot{A}^2 + g_0 C^2$. Also, the SE to be solved now reduces to the form

$$-\frac{1}{2} \psi_{x'x'} + \frac{1}{2} kx'^2 \psi + F(t') \psi = i\psi_{t'}. \tag{5.29}$$

In the second stage, another phase transformation of $\psi(x', t')$ as

$$\psi(x', t') = \exp\left[-i\int^{(t')} F(\tau)d\tau\right] \cdot \psi_1(x', t'), \tag{5.30}$$

will convert the form (5.29) to a TIDSE

$$-\frac{1}{2} \psi_{1x'x'} + \frac{1}{2} kx'^2 \psi_1 = i\psi_{1t'}, \tag{5.31}$$

for a linear oscillator in x' variable. The solution to (5.31) in terms of the Hermite polynomial $H_n(x')$ is readily known [99] and is given by

$$\psi_1(x', t') = \sum_n a_n e^{-\lambda_n t'} \cdot H_n(x'),$$

where λ_n are the eigenvalues (cf. eq. (5.3)). Finally, the solution of (5.18) for the potential (5.19) in terms of x and t can be expressed as [80]

$$\Psi(x, t) = \frac{1}{C^{1/2}} \cdot \exp\left[-i\int^{(t')} F(\tau)\, d\tau\right] . \exp[i\{\frac{1}{2}(x^2\ \dot{C}/C) - \dot{A}C\ x\}]$$

$$\cdot \sum_n a_n e^{-\lambda_n t'} \cdot H_n((x/c) + A)\ , \tag{5.32}$$

where the functions C and A can be calculated by solving (5.26) and (5.27). Interestingly, while the solution (5.32) is equivalent to the results of Truax [160], it reduces to the results of Burgan et al. [81] or those of Hartley and Ray [154] for $A = g_o = g_1 = 0$. Further applications of the above method to the generalized form of the potential, $V(x, t) = g_2(t)\, x^2 + g_1(t)\, x + g_0(t) + f((x/c) + A)$, where f is an arbitrary function, are rather straightforward.

For a detailed discussion of the wavefunction (5.32) for $n = 0$, see [165]. This method is also used to obtain an exact solution of the SE for a TD harmonic plus inverse harmonic potential (see Kaushal and Parashar [201]) of the form $V(x, t) = a_2(t)\,x^2 + a_1/x^2$.

5.3 FEYNMAN PROPAGATOR APPROACH

An alternative way of quantizing the system is through the Feynman propagator $K(x'', t''; x', t')$ defined as the path integral [161]

$$K(x'', t''; x', t') = \int \exp\left(i \int_{t'}^{t''} L dt . \right) \mathcal{D}(x(t)), \tag{5.33}$$

where L is the Lagrangian and integrations are over all paths starting at $x' = x(t')$ and terminating at $x'' = x(t'')$. It is interesting to note that for a quadratic Lagrangian the propagator takes the form [162]

$$K(x'', t''; x', t') = F(t'', t').\exp\,[i\, S_{cl}\,(x'', t''; x', t')/\hbar], \tag{5.34}$$

where S_{cl} is the classical action and $F(t'',t')$ is an entirely TD function. (For $t' = 0$ and $t'' = t$, F is a function of t alone.) In the past, the function $F(t'', t')$ has been obtained using different methods. Earlier, Feynman and Hibbs [162] described $F(t'', t')$ as a conditional path integral and it is given by the van Vleck-Pauli formula [163] as $[(i/2\pi)\cdot\partial^2 S_{cl}\,(x'', t''; x', t')/\partial x''\cdot\partial x']^{1/2}$. Later, $F(t'', t')$ was evaluated by Goovaerts [164] taking recourse to the SE (5.1) and by Papadopoulos [164] for a generalized quadratic Langrangian. Khandekar and Lawande [159] obtained it directly from the TDSE by interpreting $F(t'', t')$ as the limit of multiple Riemann integrals. In fact, this latter approach while dealing with $F(t'', t')$ as the normalizing factor, accounts also for the contribution due to the classical path simultaneously.

An important relation which exists (cf. Sect. 5.2) between the LR [64] theory in terms of the invariant operator $I(t)$ and the Feynman propagator can be deduced as follows: From (5.2) the expansion coefficients c_n can be obtained as

$$c_n = \exp(-i\alpha_n(t)) . \int \psi_n^*(x, t)\; \Psi(x, t)\, dx. \tag{5.35}$$

Using this result the wavefunction Ψ at the position x'' and time t'' can be expressed as

$$\Psi(x'', t'') = \int dx\; \exp\,[i\{\alpha_n(t'') - \alpha_n(t')\}]\cdot\psi_n^*(x', t')\;\psi_n(x'', t'') \cdot \Psi(x', t'). \tag{5.36}$$

After comparing this expression with the definition of the propagator, viz,

$$\Psi(x'',t'') = \int K(x'',t'';x',t')\ \Psi(x',t')\ dx',\ (t'' > t'), \quad (5.37)$$

one immediately obtains

$$K(x'',t'';x',t') = \sum_n \exp[i(\alpha_n(t'') - \alpha_n(t'))] \cdot \psi_n^*(x',t')\ \psi_n(x'',t''), \quad (5.38)$$

which represents usual expansion formula for the TD Hamiltonian given by Feynman and Hibbs [162]. Thus, the existence of the invariant $I(t)$ through its eigenfunction $\psi_n(x, t)$ (cf.eq. (5.3)) considerably simplifies the derivation of the propagator.

The Feynman propagator approach using the path integral method while successfully works [19] for the TID systems, its applications to TD systems have several advantages. In fact, for a given Lagrangian not only the quantum mechanical considerations may be introduced more directly through this approach but the quantum superposition is also built-in in this method. Also, if (5.34) is correct then the classical limit of quantum mechanics can be understood with little mathematical effort, of course, without actually going into the details of semiclassical physics. Further, the nature of the TD invariant (particularly its dependence on the powers of p) is also crucial as far as the applications of this approach are concerned. A number of TD systems of somewhat complicated structure but admitting second-order (in momenta) invariants are investigated by Khandekar and Lawande [159] within this framework. Though for the details of 1D systems we refer to their review [19], the study of TD systems in 2D has not yet been carried out using this approach. Finally, it may be mentioned that the propagator theory in general and the exact solutions for various TD phenomena have also been studied [165] in other quantum context. However, we refrain ourselves from such discussions here.

5.4 EXACT SOLUTION TO TDSE IN 2D

Here we first discuss the generalization of the method of Ray [80] in 2D and then make some remarks on the results of Burgan et al. [81] obtained in higher dimensions. In particular, now we consider the exact solution of the TDSE ($\mu = \hbar = 1$)

$$[-\frac{1}{2}\{(\partial^2/dx_1^2) + (\partial^2/dx_2^2)\} + V(x_1,x_2,t)]\ \Psi(x_1,x_2,t) = i(\partial\Psi/\partial t), \quad (5.39)$$

by making an ansatz for the function $\Psi(x_1, x_2, t)$ as [166]

$$\Psi(x_1, x_2, t) = B(t).\exp[i\phi(x_1, x_2, t)].\ \psi(x_1, x_2, t). \quad (5.40)$$

Here $B(t)$ plays the same role as in (5.20) and for other space and time variables now we use the following scale transformation

$$x_1' = \frac{x}{C_1(t)} + A_1(t) \; ; \; x_2' = \frac{x_2}{C_2(t)} + A_2(t) \; ; \; t' = D(t), \quad (5.41)$$

where $A_i(t)$, $C_i(t)$ and $D(t)$ are the arbitrary functions of t as before. Corresponding to the expression (5.22), now we have

$$-\frac{1}{2}\left[iB\phi_{11}\psi - B\phi_1^2\psi + 2i\frac{B}{C_1}\phi_1\psi_{1'} + \frac{B}{C_1^2}\psi_{1'1'} + iB\phi_{22}\psi - B\phi_2^2\psi \right.$$

$$\left. + 2i\frac{B}{C_2}\phi_2\psi_{2'} + \frac{B}{C_2^2}\psi_{2'2'} \right] + VB\psi$$

$$= i\dot{B}\psi - B\psi_t\psi + iB\dot{D}\psi_{t'} - iB[\frac{\dot{C}_1}{C_1^2}x_1 - \dot{A}_1]\psi_1 - iB[\frac{\dot{C}_2}{C_2^2}x_2 - \dot{A}_2]\psi_2, \quad (5.42)$$

where subscripts to ϕ and ψ indicate the partial derivatives of these functions in the spirit of eq. (4.32) including w.r.t. the primed variables. The expression for the function ϕ now turns out to be

$$\phi = \frac{1}{2}\left[\frac{\dot{C}_1}{C_1}x_1^2 + \frac{\dot{C}_2}{C_2}x_2^2\right] - (\dot{A}_1 C_1 x_1 + \dot{A}_2 C_2 x_2). \quad (5.43)$$

Using (5.43) in (5.42) and subsequently demanding the invariance of the form of the SE w.r.t. the primed and unprimed variables, one can fix some of the arbitrary TD functions as [166]

$C_1(t) = C_2(t) = C(t)$, $t' \equiv D(t) = \int dt/C^2(t)$; $B(t) = 1/C(t)$.

As a result, (5.43) also takes the simpler form,

$$\phi = \frac{\dot{C}}{2C}(x_1^2 + x_2^2) - C(\dot{A}_1 x_1 + \dot{A}_2 x_2). \quad (5.43a)$$

Finally, eq. (5.42) reduces to the form

$$-\frac{1}{2}(\psi_{1'1'} + \psi_{2'2'}) + V'(x_1', x_2', t') = i\psi_{t'}, \quad (5.44)$$

where the potential V' is now given by

$$V' = VC^2 + \frac{1}{2}C\ddot{C}(x_1^2 + x_2^2) - C^2(\ddot{A}_1 C + 2\dot{A}_1\dot{C})x_1 - C^2(\ddot{A}_2 C + 2\dot{A}_2\dot{C})x_2$$

$$+\frac{1}{2}\, C^4 (\dot{A}_1^2 + \dot{A}_2^2). \tag{5.45}$$

Note the difference in the expression for $B(t)$ here as compared to that for the case of 1D. Further, by setting $x_2 = 0$ and $A_2(t) = 0$ in (5.41) and (5.45) one can easily recover the results of Ray [80] for 1D. If, in addition, $A_1(t) = 0$, then one can arrive at the results of Burgan et al. [81]. In this respect the multidimensional studies of Burgan et al. are of highly restricted nature. Next, we demonstrate some applications of these results to NC harmonic and NCAH, TD potentials.

5.4.1 Example of a NC Harmonic Potential

Burgan et al. [81], have studied a TD multidimensional potential which is just a superposition of harmonic potentials along each dimension. Here we consider a simple generalization of the potential (5.19) in the form

$$V(x_1, x_2, t) = b_{2o}(t)\, x_1^2 + b_{02}(t)\, x_2^2 + b_{11}(t)\, x_1 x_2 + b_{10}(t)\, x_1 + b_{01}(t)\, x_2 + b_o(t), \tag{5.46}$$

which, for $b_{11} = 0$, is just the case of a shifted rotating harmonic oscillator. We analyse this case in detail. Using the inverse of the transformation (5.41), the potential $V'(x_1', x_2', t')$ can be computed from (5.45) as

$$\begin{aligned} V' &= C^3 (b_{2o}\, C + \frac{1}{2}\, \ddot{C})\, x_1'^2 + C^3 (b_{02} C + \frac{1}{2}\, \ddot{C})\, x_2'^2 \\ &\quad + C^3 [(b_{10} - \ddot{A}_1\, C - 2\dot{A}_1\, \dot{C}) \\ &\quad - 2A_1 (b_{20}\, C + \frac{1}{2}\, \ddot{C})]\, x_1' + C^3 [(b_{01} - \ddot{A}_2\, C - 2\dot{A}_2 \dot{C}) \\ &\quad - 2A_2 (b_{02} C + \frac{1}{2}\ddot{C})]\, x_2' + F(t'), \end{aligned} \tag{5.47}$$

where

$$\begin{aligned} F(t') &= A_1 C^3 [A_1 (b_{20} C + \frac{1}{2}\ddot{C}) - (b_{10} - \ddot{A}_1\, C - 2\dot{A}_1\, \dot{C})] \\ &\quad + A_2\, C^3 [A_2 (b_{02} C + \frac{1}{2}\ddot{C}) - (b_{01} - \ddot{A}_2\, C - 2\dot{A}_2\, \dot{C})] + b_0 C^2 + \frac{1}{2} C^4 (\dot{A}_1^2 + \dot{A}_2^2). \end{aligned} \tag{5.48}$$

Let the function $C(t)$ satisfy the equations

$$\ddot{C} + 2b_{20}\, C = k_1\, C^{-3};\; \ddot{C} + 2b_{02}\, C = k_2\, C^{-3}. \tag{5.49}$$

From (5.49), $C(t)$ can be determined in terms of the potential parameters b_{02} and b_{20} and the constants k_1 and k_2. In order to determine $A_1(t)$ and $A_2(t)$, we equate to zero the coefficients of x_1' and x_2' in (5.47), leading to

$$\ddot{A}_1 + \frac{2\dot{A}_1\dot{C}}{C} + \frac{A_1 k_1}{C^4} - \frac{b_{10}}{C} = 0;\; \ddot{A}_2 + \frac{2\dot{A}_2\dot{C}}{C} + \frac{A_2 k_2}{C^4} - \frac{b_{01}}{C} = 0. \tag{5.50}$$

Thus, the expression (5.47) for V' takes the form

$$V' = \frac{1}{2}\, k_1\, x_1'^2 + \frac{1}{2}\, k_2\, x_2'^2 + F\,(t'), \tag{5.51}$$

with

$$F(t') = -\frac{1}{2} k_1\, A_1^2 - \frac{1}{2} k_2\, A_2^2 + \frac{1}{2}\, C^4\,(\dot{A}_1^2 + \dot{A}_2^2) + b_0\, C^2. \tag{5.52}$$

Finally, the exact solution to eq. (5.39) for the potential (5.46) with $b_{11}\,(t) = 0$, can be written in analogy with eq. (5.32) as [166]

$$\Psi\,(x_1, x_2, t) = \frac{1}{2} \exp[-i\int^{(t')} F(\tau)d\tau].\exp[\frac{i}{2C}\{\dot{C}(x_1^2 + x_2^2) - C^2(\dot{A}_1 x_1 + \dot{A}_2 x_2)\}]$$

$$\cdot \sum_m \sum_n \exp[-i(\lambda_m + \lambda_n) \int dt / C^2] \cdot H_m(x / C + A_1).H_n(x_2 / C + A_2), \tag{5.53}$$

where $C(t)$, $A_i(t)$, and $F(t)$, respectively can be obtained from eqs. (5.49), (5.50) and (5.52).

Note that if the coupling term $b_{11} \neq 0$ in the potential (5.46), then a term of the type $c^4 b_{11} x_1'\, x_2'$ also appears in the expression for V' (cf.eq. (5.51)) and hence does not allow the TDSE to reduce to a TID form of the SE even after carrying out the second stage phase transformation of the type (5.30). Unfortunately, such a reduction from TD problem to a TID one is possible only for a limited class of NC harmonic potentials [166] and for AH potentials too it is not an easy task.

5.4.2 Case of Anharmonic Potentials

The limitations of the present method for an AH TD potential of the form

$$\begin{aligned} V(x_1, x_2, t) \;=\; & b_{40} x_1^4 + b_{04} x_2^4 + b_{22} x_1^2 x_2^2 + b_{20} x_1^2 + b_{02} x_2^2 \\ & + b_{11} x_1 x_2 + b_{10} x_1 + b_{01} x_2 + b_0, \end{aligned} \tag{5.54}$$

where b_o and b_{ij}'s are the functions of t, can be demonstrated as follows. In this case the potential V' from (5.45), after using the inverse of the transformation (5.41), becomes

$$\begin{aligned}
V'(x_1', x_2', t') = {} & C^6 (b_{40} x_1'^4 + b_{04} x_2'^4 + b_{22} x_1'^2 x_2'^2 - 4 b_{40} A_1 x_2'^3 - 4 b_{04} A_2 x_2'^3 \\
& -2 b_{22} A_1 x_1' x_2'^2 - 2 b_{22} A_2 x_1'^2 x_2') + C^4 (4C^2 b_{22} A_1 A_2 + b_{11}) x_1' x_2' + \\
& + C^3 (6C^3 b_{40} A_1^2 + C^3 b_{22} A_2^2 + C b_{20} + \frac{1}{2}\ddot{C}) x_1'^2 + C^3 (6C^3 b_{04} A_2^2 \\
& + C^3 b_{22} A_1^2 + C b_{02} + \frac{1}{2}\ddot{C}) x_2'^2 \\
& - C^3 (2C^3 b_{40} A_1^3 + 2C^3 b_{22} A_1 A_2^2 + C b_{11} A_2 + 2C b_{20} A_1 + A_1 \ddot{C} \\
& + C\ddot{A}_1 + 2\dot{A}_1 \dot{C} - b_{10}) x_1' \\
& - C^3 (2C^3 b_{04} A_2^3 + 2C^3 b_{22} A_1^2 A_2 + C b_{11} A_1 + 2C b_{02} A_2 + A_2 \ddot{C} \\
& + C\ddot{A}_2 + 2\dot{A}_2 \dot{C} - b_{01}) x_2' + F(t'),
\end{aligned} \tag{5.55}$$

with

$$\begin{aligned}
F(t') = {} & C^3 [C^3 b_{40} A_1^4 + C^3 b_{04} A_2^4 + C^3 b_{22} A_1^2 A_2^2 + C b_{11} A_1 A_2 + A_1^2 \\
& (C b_{20} + \frac{1}{2}\ddot{C}) + A_2^2 (C b_{02} + \frac{1}{2}\ddot{C}) + A_1 (C\ddot{A}_1 + 2\dot{A}_1 \dot{C} - b_{10}) + A_2 \\
& (C\ddot{A}_2 + 2\dot{A}_2 \dot{C} - b_{01}) + \frac{1}{2} C (\dot{A}_1^2 + \dot{A}_2^2) - C b_{11} A_1 A_2] + b_0 C^2 .
\end{aligned}$$

Since the potential now necessarily contains anharmonic terms, it is not possible to reduce the TDSE (5.39) with the potential (5.54) to a TIDSE with a harmonic potential unless the transformation (5.41) in space variables becomes nonlinear. This will, however, lead to other complications in applying the present method. Alternatively, one can assume the coefficients of the AH terms in $V(x_1, x_2, t)$ to be TID and the time-dependence mainly appears in the quadratic and linear terms only. In this case, however, one will need the exact solution to the TIDSE with the corresponding AH potential and such a solution is not often available in the literature except for a few limited cases [97]. Thus, as far as the applications of the present method to nonharmonic TD potentials in general are concerned, they are not as transparent as for the harmonic potentials.

5.5 CONCLUDING DISCUSSION

From the point of view of having an exact solution of the problem the scope of study of quantum mechanics of NC TD systems, not only in this Chapter but also in general in the literature itself, has been highly limited as compared to that of NC TID systems. This is mainly for two reasons. Firstly, because in spite of several experimentally tenable phenomena (e.g. in the field of femtochemistry [217]) in two or higher dimensions which can manifest the time-dependence (if at all it exists) of the underlying interactions, their theoretical study in terms of exact solutions is still inadequate. Secondly, the available techniques developed primarily for the 1D systems and they appear to be inadequate even for these systems, are not easily extendable to two or higher dimensional TD systems. Moreover, in this Chapter our aim has been to look only into the exact solutions of the TDSE which correspond to the complete spectrum of eigenvalues. In particular, the possibility of using the eigenfunction-ansatz method which is based on the LR [64] theory and carries the spirit of dynamical invariants, is demonstrated for the 2D systems (cf. Sect. 5.4). This method, while seems to work well for TD harmonic potentials, again shows limitations for TD anharmonic or other nonpolynomial potentials.

As far as the development of approximation methods for TD systems in two and higher dimensions is concerned that can still be awaited until the applicability of these methods can completely be understood for (i) TD systems in 1D and for (ii) TID systems in two and higher dimensions [141] (cf. Sect. 4.5). For the TD case even in 1D, besides the perturbation method the applicability of other conventional methods (like variational or WKB methods) have not yet been completely understood (see, e.g. Kleber [165]). Only recently (see, e.g. Breuer and Holthaus [165], Bensch et al. [165] and Mirbach and Korsch [165]) there have been some attempts to develope EBK-type (cf. Sect. 4.5) quantization condition for time-periodic Hamiltonians in 1D. In fact, in some physical problems the combination of the two (variational and perturbation) methods turns out to be useful.

No doubt the methods which deal with somewhat more mathematical rigour (like Lie algebraic or Feynman propagator approach discussed, respectively in Sects. 5.2.2 and 5.3) require further exploration as far as the study of higher-dimensional systems is concerned but even in their present form in 1D the applicability for Hamiltonians beyond the quadratic form is limited. It may be mentioned that the method employed in Sects. 5.2.3 and 5.4 essentially has a basis in the simple theorem stated in Sect. 5.2.2. This theorem, to some extent sets the criterion for the solvability of a TD system. In fact, according to this theorem $H(t)$ always admits a TD invariant operator

$I(t)$ due to the arbitrariness of $I(0)$ provided the evolution operator $U(t)$ for the given system can be constructed easily. However if one looks for a constant of motion (i.e. for a nontrivial TID invariant $\mathcal{I}$), then eq. (5.15) on differentiation would demand $[\mathcal{I}, H(t)] = 0$. Further, if there is no such $I(t)$ for a given $H(t)$, then the problem does not admit any solution.

6

Noncentral Forces in Three Dimensions: A Brief Survey

As far as the study of NC systems in 3D (whether at the classical or quantum level) is concerned, not many efforts have been made in the literature except for the possibilities of extending the methods and techniques developed already for 1D or 2D systems. Even in this case the central force problems investigated in 3D do not lie within the purview of this chapter and we avoid their discussion. Therefore, we shall make a few remarks (more or less in a passing manner) on the works carried out on the NC potentials in 3D, discussing both classical and quantum aspects. Although it appears trivial to extend the methods and techniques available already for the 2D systems to the case of 3D systems nevertheless it is necessary to highlight the underlying difficulties in this process. In fact, this is also essential from the point of view of applications of these methods to physical problems in 3D which are perhaps more realistic and occur more frequently than those in 2D. In the next Section, we discuss the classical mechanical aspect of some 3D systems and in Sect. 6.2 the quantum aspect will only be touched upon as sufficient literature is not available on this subject. Some concluding remarks are made in Sect. 6.3.

6.1 CLASSICAL ASPECT

In this context we again classify the discussion on the basis of explicit time-dependence of 3D systems.

6.1.1 TID Systems in 3D

As compared to the 2D case the study of 3D systems turns out to be somewhat more involved. In what follows we briefly outline the extension of some of the methods used (cf. Sects. 2.3.1 and 2.3.3) for the construction of invariants for 2D systems, to the case of 3D systems.

(a) Rationalization method: In analogy with the 2D case (cf. Sect. 2.3.1) the rationalization method can also be applied to the 3D systems described by the Lagrangian

$$L = \frac{1}{2}(\dot{x}_1^2 + \dot{x}_2^2 + \dot{x}_3^2) - V(x_1, x_2, x_3), \tag{6.1}$$

with the corresponding equations of motion

$$\ddot{x}_i = (\partial L / \partial x_i) \equiv \alpha_i(x_1, x_2, x_3), \ (i = 1, 2, 3). \tag{6.2}$$

We demonstrate here the derivation only of linear and quadratic (in momenta) invariants. For the linear invariant of the form $I_1 = a_i \xi_i$, eqs. (2.12) and (2.14) in the Cartesian coordinates will yield ($i, j = 1, 2, 3$)

$$(\partial a_1/\partial x_1) = 0; \ (\partial a_2/\partial x_2) = 0; \ (\partial a_3/\partial x_3) = 0; \ (\partial a_1/\partial x_2) + (\partial a_2/\partial x_1) = 0,$$
$$(\partial a_1/\partial x_3) + (\partial a_3/\partial x_1) = 0 \ ; \ (\partial a_2/\partial x_3) + (\partial a_3/\partial x_2) = 0, \tag{6.3}$$
$$a_1(\partial V/\partial x_1) + a_2(\partial V/\partial x_2) + a_3(\partial V/\partial x_3) = 0.$$

Noting the fact that first three equations above imply

$$a_1 = f_1(x_2, x_3) \ ; \quad a_2 = f_2(x_1, x_3) \ ; \quad a_3 = f_3(x_1, x_2),$$

use of the next three equations in turn will lead to the expressions for a_i's as

$$a_1 = c_1 x_2 - c_4 x_3 + c_2' \ ; \ a_2 = c_7 x_3 - c_1 x_1 + c_3' \ ; \ a_3 = c_4 x_1 - c_7 x_2 + c_5',$$

with $c_2' = c_2 + c_6$; $c_3' = c_3 + c_8$; $c_5' = c_5 + c_9$. Here c_1, c_4 are c_7 are the separation constants and other c_i's are the constants of integration. These constants can be obtained by rationalizing the last equation in (6.3) for a given V and thus finally providing the linear invariant I_1.

For the quadratic invariant of the form $I_2 = a_0 + a_{ij} \xi_i \xi_j$, eqs. (2.11) and (2.13) will now yield ($i, j = 1, 2, 3$) the following 13 equations compared to 6 equations for the 2D case:

$$(\partial a_{11}/\partial x_1) = 0; \ (\partial a_{22}/\partial x_2) = 0; \ (\partial a_{33}/\partial x_3) = 0; \ (\partial a_{11}/\partial x_2) + 2(\partial a_{12}/\partial x_1) = 0$$

(6.4a, b, c, d)

$$(\partial a_{11}/\partial x_3) + 2(\partial a_{13}/\partial x_1) = 0; \quad (\partial a_{22}/\partial x_1) + 2(\partial a_{12}/\partial x_2) = 0;$$

$$(\partial a_{33}/\partial x_1) + 2(\partial a_{13}/\partial x_3) = 0, \tag{6.4e, f, g}$$

$$(\partial a_{22}/\partial x_3) + 2(\partial a_{23}/\partial x_2) = 0; \ (\partial a_{33}/\partial x_2) + 2(\partial a_{23}/\partial x_3) = 0, \tag{6.4h, i}$$

$$(\partial a_{12}/\partial x_3) + (\partial a_{23}/\partial x_1) + (\partial a_{31}/\partial x_2) = 0, \tag{6.4j}$$

$$(\partial a_0/\partial x_1) = a_{11}(\partial V/\partial x_1) + a_{12}(\partial V/\partial x_2) + a_{13}(\partial V/\partial x_3), \quad (6.4k)$$

$$(\partial a_0/\partial x_2) = a_{12}(\partial V/\partial x_1) + a_{22}(\partial V/\partial x_2) + a_{23}(\partial V/\partial x_3), \quad (6.4l)$$

$$(\partial a_0/\partial x_3) = a_{13}(\partial V/\partial x_1) + a_{23}(\partial V/\partial x_2) + a_{33}(\partial V/\partial x_3). \quad (6.4m)$$

First three of these equations immediately give

$$a_{11} = a_{11}(x_2, x_3), \quad a_{22} = a_{22}(x_1, x_3), \quad a_{33} = a_{33}(x_1, x_2),$$

which, are when used in (6.4 d) and (6.4 e), lead to two expressions for a_{12} as

$$a_{12} = h_1(x_2, x_3)x_1 + h_2(x_2, x_3), \; a_{12} = h_3(x_1, x_3)x_2 + h_4(x_1, x_3),$$

where h_i's are arbitrary functions of their arguments. Noting that these two expressions for a_{12} have to be the same, one has to restrict to the choices of h_i's to

$$h_1 = s_3x_2 + c, \; h_2 = cx_2 + k, \; h_3 = s_3x_1 + c, \; h_4 = cx_1 + k,$$

which leads to

$$a_{12} = s_3\, x_1x_2 + c\,(x_1 + x_2) + k. \quad (6.5a)$$

Here s_3, c and k are constants. Similarly, one can obtain

$$a_{13} = s_2x_1x_3 + c\,(x_1 + x_3) + k, \; a_{23} = s_1x_2x_3 + c\,(x_2+x_3) + k, \quad (6.5b,c)$$

Use of these results for a_{12}, a_{13} and a_{23} in eqs. (6.4d) - (6.4i), will give the expressions for a_{11}, a_{22} and a_{33} in the form as [167]

$$a_{11} = -\, s_3\, x_2^2 - s_2\, x_3^2 - 2c\,(x_2 + x_3) + k_1, \quad (6.5d)$$

$$a_{22} = -\, s_3\, x_1^2 - s_1\, x_3^2 - 2c\,(x_1 + x_3) + k_2, \quad (6.5e)$$

$$a_{33} = -\, s_2\, x_1^2 - s_1\, x_2^2 - 2c\,(x_1 + x_2) + k_3, \quad (6.5f)$$

where k_i's are constants.

Now, the elimination of a_0 from the pairs of eqs. [(6.4k), (6.4l)], [(6.4l), (6.4m)] and [(6.4k), (6.4m)] will respectively lead to the following three potential equations unlike the single potential equation (2.2) in the 2D case

$$[(\partial a_{11}/\partial x_2) - (\partial a_{12}/\partial x_1)]\,(\partial V/\partial x_1) + [(\partial a_{12}/\partial x_2) - (\partial a_{22}/\partial x_1)\,]\,(\partial V/\partial x_2) +$$
$$[(\partial a_{13}/\partial x_2) - (\partial a_{23}/\partial x_1)]\,(\partial V/\partial x_3) + (a_{11} - a_{22})\,(\partial^2 V/\partial x_1\partial x_2) +$$
$$a_{12}\,[(\partial^2 V/\partial x_2^2) - (\partial^2 V/\partial x_1^2)] + [a_{13}\,(\partial^2 V/\partial x_2\partial x_3) - a_{23}\,(\partial^2 V/\partial x_1\partial x_3)] = 0, \quad (6.6a)$$

$$[(\partial a_{22}/\partial x_3) - (\partial a_{23}/\partial x_2)]\,(\partial V/ax_2) + [(\partial a_{23}/\partial x_3) - (\partial a_{33}/\partial x_2)]\,(\partial V/\partial x_3) +$$

$$[(\partial a_{12}/\partial x_3) - (\partial a_{13}/\partial x_2)]\,(\partial V/\partial x_1) + (a_{22} - a_{33})\,(\partial^2 V/\partial x_2\partial x_3) + a_{23}\,[(\partial^2 V/\partial x_3^2) -$$

$$(\partial^2 V/\partial x_2^2)] + [a_{12}\,(\partial^2 V/\partial x_3\partial x_1) - a_{13}\,(\partial^2 V/\partial x_2\,\partial x_1)] = 0, \qquad (6.6b)$$

$$[(\partial a_{33}/\partial x_1) - (\partial a_{13}/\partial x_3)]\,(\partial V/\partial x_3) + [(\partial a_{13}/\partial x_1) - (\partial a_{11}/\partial x_3)]\,(\partial V/\partial x_1) +$$

$$[(\partial a_{23}/\partial x_1) - (\partial a_{12}/\partial x_3)]\,(\partial V/\partial x_2) + (a_{33} - a_{11})\,(\partial^2 V/\partial x_3\partial x_1) +$$

$$a_{13}\,[(\partial^2 V/\partial x_1^2) + (\partial^2 V/\partial x_3^2)] + [a_{23}\,(\partial^2 V/\partial x_1\partial x_2) - a_{12}\,(\partial^2 V/\partial x_3\,\partial x_2)] = 0. \qquad (6.6c)$$

For a given V one has to rationalize all the three potential equations (6.6a)-(6.6c) and determine some or all of the arbitrary constants contained in a_{ij}'s. Then, in the end one has to rationalize the eqs. (6.4k) - (6.4m) not only to determine the remaining arbitrary constants but also to obtain a unique expression for a_o. Once the coefficient functions a_o and a_{ij}'s become known, the invariant I_2 can be obtained as before. Using the above prescription Mishra and Parashar [167a] have constructed at least one invariant for a number of 3D systems.

(*b*) *Method of Lie symmetries*: Here, following the work of Lakshmanan and Sahadevan [46] as outlined in Sect. 2.3.3, we continue to discuss Lie symmetries and associated integrals of motion for the 3D systems described by the Lagrangian (6.1).

For this purpose we consider the invariance of the equations of motion (6.2), under the one-parameter ($\in$) group of transformations

$$t \to T = T\,(t, x_1, x_2, x_3, \dot{x}_1, \dot{x}_2, \dot{x}_3, \in),$$

$$x_i \to X_i = X_i\,(t, x_1, x_2, x_3, \dot{x}_1, \dot{x}_2, \dot{x}_3, \in), \qquad (6.7)$$

with the associated forms $\dot{X}_i = \partial X_i/\partial T$, $i = 1, 2, 3$. The corresponding infinitesimal transformations for small $\in$ ($\in << 1$) are now given by

$$t \to T = t + \in \xi\,(t, x_1, x_2, x_3, \dot{x}_1, \dot{x}_2, \dot{x}_3) + 0\,(\in^2),$$

$$x_i \to X_i = x_i + \in \eta_i\,(t, x_1, x_2, x_3, \dot{x}_1, \dot{x}_2, \dot{x}_3) + 0\,(\in^2), \qquad (6.8)$$

$$\dot{x}_i \to \dot{X}_i = \dot{x}_i + \in (\dot{\eta}_i - \dot{\xi}\,\dot{x}_i),$$

where $\dot{\eta}_i = \Gamma\,\eta_i$, $(i = 1, 2, 3)$; $\dot{\xi} = \Gamma\,\xi$, and

$$\Gamma = \frac{\partial}{\partial t} + \sum_{i=1}^{3}\left(\frac{\partial}{\partial x_i}\dot{x}_i + \alpha_i\frac{\partial}{\partial \dot{x}_i}\right).$$

As a result, the invariance condition for the equations of motion leads to

$$\ddot{\eta}_i - \dot{x}_i \ddot{\xi} - 2\dot{\xi}\alpha_i = E(\alpha_i), \ (i = 1, 2, 3), \tag{6.9}$$

where

$$E = \xi \frac{\partial}{\partial t} + \sum_{i=1}^{3} \left(\eta_i \frac{\partial}{\partial x_i} + (\dot{\eta}_i - \dot{\xi}\,\dot{x}_i) \frac{\partial}{\partial \dot{x}_i} \right). \tag{6.10}$$

To determine nontrivial forms of the infinitesimal symmetries η_1, η_2, η_3 and ξ one makes ansatze for them in terms of polynomials in the velocities $\dot{x}_1, \dot{x}_2, \dot{x}_3$ as before and subsequently obtain an expression for the associated conserved quantity I (if it exists) as

$$I = \sum_{i=1}^{3} (\xi\,\dot{x}_i - \eta_i) \frac{\partial L}{\partial \dot{x}_i} - \xi L + f, \tag{6.11}$$

where the function $f(x_1, x_2, x_3)$ is to be determined from

$$E\{L\} + \dot{\xi} L = \sum_{i=1}^{3} \frac{\partial f}{\partial x_i} \dot{x}_i, \tag{6.12}$$

so that I is an invariant. For the case when $\eta_1 = \eta_2 = \eta_3 = 0$, ξ = const., the function $f(x_1, x_2, x_3)$ turns out to be the potential $V(x_1, x_2, x_3)$ as before and one out of the three invariants (since in this case the system, if integrable, should admit three invariants following Whittaker [9]) can be identified with the Hamiltonian. It may be mentioned that a nontrivial determination of the symmetries in this case would, however, require the solutions of PDE's whose number is now an order of magnitude higher than the case of a corresponding 2D system. Further, the symmetries are assumed to be TID; otherwise the problem would become more complicated. Lakshmanan and Sahadevan [46] have used these results to study the cases of three-coupled, quartic and cubic AH oscillators described respectively by the Lagrangians

$$L = \frac{1}{2}(\dot{x}_1^2 + \dot{x}_2^2 + \dot{x}_3^2) - [\, b_{200}x_1^2 + b_{020}x_2^2 + b_{002}x_3^2 + b_{400}x_1^4 + b_{040}x_2^4 + b_{004}x_3^4 + b_{220}x_1^2\,x_2^2 + b_{022}x_2^2\,x_3^2 + b_{202}x_1^2\,x_3^2\,], \tag{6.13}$$

and

$$L = \frac{1}{2}(\dot{x}_1^2 + \dot{x}_2^2 + \dot{x}_3^2) - [\frac{1}{2}(b_{200}x_1^2 + b_{020}x_2^2 + b_{002}x_3^2) +$$

$$+ b_{210}x_1^2 x_2 + b_{012}x_2 x_3^2 - \frac{1}{3} b_{030}x_2^3] , \tag{6.14}$$

where b_{ijk}'s are the TID parameters.

In a series of papers [167b] Lakshmanan and Sahadevan have also carried out the Painleve' analysis and made an attempt to identify the integrable cases of three-coupled nonlinear oscillators. In particular, the cases of quartic oscillators described by the Lagrangian (6.13) and of sextic oscillators corresponding to the potential,

$$V(x_1, x_2, x_3) = b_{200}x_1^2 + b_{020}x_2^2 + b_{002}x_3^2 + b_{600}x_1^6 + b_{060}x_2^6 + b_{006}x_3^6 +$$

$$b_{420}x_1^4 x_2^2 + b_{402}x_1^4 x_3^2 + b_{042}x_2^4 x_3^2 + b_{240}x_1^2 x_2^4 + b_{204}x_1^2 x_3^4 +$$

$$b_{024}x_2^2 x_3^4 + b_{222}x_1^2 x_2^2 x_3^2 , \tag{6.15}$$

are investigated. The Painleve' analysis of N-coupled anharmonic oscillators in higher dimensions is also carried out [18]. It may be recalled (cf. Sect. 2.3.2) that the Painleve' method has limitations as far as the construction of exact invariants is concerned. For the details of this method we again refer to the literature cited in Ref. [18].

The situation of a 3D system can also arise in the study of a four-particle linear chain in the same way as 2D systems (like Toda potential etc.) are obtained [34, 47] from a three-particle linear chain. Such systems in general have nonlinear constants of motion. Recently, Ranada [47] has investigated these constants of motion for an integrable three-particle system using the theory of generalized symmetries. Another TID system in 3D, well known in the literature for a long time, is the case of an asymmetric top rotating around a fixed point under the influence of its own weight. However, we postpone the discussion of the integrability (both classical and quantal) of such a system to Appendix 2.

6.1.2 TD Systems in 3D: Dynamical Algebraic Approach

As mentioned before (cf. Chap. 3) that the explicit dependence on time of a dynamical system further complicates its study, the same is true for the TD systems in 3D. In fact, the extension of the rationalization method from 1D to 2D, discussed already in Chap. 3, can as well be carried out for the present 3D case but an increase in the degree of complexity can easily be visualized even for the second-order invariants. With a view to demonstrating the magnitude of such complexity we present here a possible extension of the Lie algebraic approach of Sects. 3.2.3 and 3.3.2 to the 3D case.

For a TD system in 3D, the Hamiltonian (in analogy with eq. (3.65)) can be expressed as

$$H = \sum_n h_n (t) \Gamma_n (x_1, p_1, x_2, p_2, x_3, p_3), \tag{6.16}$$

where Γ_n's are the phase space functions of the arguments and they are required to close the algebra as before with respect to the Poisson bracket (3.66) but is now defined for the 3D case. The invariant I, now expressed as

$$I = \sum_k \lambda_k (t) \Gamma_k (x_1, p_1, x_2, p_2, x_3, p_3), \tag{6.17}$$

also fulfils the requirement (3.25) which leads to a similar set of equations as (3.26) for obtaining the unknown λ_k's of (6.17). To be specific, we consider the case of three-coupled TD oscillators described by the Hamiltonian,

$$H = \frac{1}{2}(p_1^2 + p_2^2 + p_3^2) + \alpha_1 (t) x_1^2 + \alpha_2 (t) x_2^2 + \alpha_3 (t) x_3^2 +$$

$$\beta (t) \phi (x_1, x_2, x_3). \tag{6.18}$$

For this case, the Γ_i's of (6.16) can be chosen as

$$\Gamma_1 = (p_1^2 / 2); \Gamma_2 = (p_2^2 / 2); \Gamma_3 = (p_3^2 / 2); \Gamma_4 = x_1^2;$$

$$\Gamma_5 = x_2^2; \Gamma_6 = x_3^2; \Gamma_7 = \phi,$$

and the corresponding h_i's as

$$h_1 = h_2 = h_3 = 1,\ h_4 = \alpha_1(t),\ h_5 = \alpha_2(t),\ h_6 = \alpha_3(t),\ h_7 = \beta(t).$$

In order to close the algebra in this case one needs to introduce three more Γ_i's, namely $\Gamma_8 = -2p_1x_1$, $\Gamma_9 = -2p_2x_2$, $\Gamma_{10} = -2p_3x_3$, with corresponding h_i's as $h_8 = h_9 = h_{10} = 0$. Note that the number of nonvanishing Poisson brackets to be computed to close the algebra turns out to be much larger in this case as compared to that of the 2D case. Due to the presence of the ϕ-term in (6.18), the affected (nonvanishing) Poisson brackets are

$$[\Gamma_1, \Gamma_7]_{PB} = -p_1 (\partial\phi / \partial x_1); [\Gamma_2, \Gamma_7]_{PB} = -p_2 (\partial\phi / \partial x_2);$$

$$[\Gamma_3, \Gamma_7]_{PB} = -p_3 (\partial\phi / \partial x_3); [\Gamma_7, \Gamma_8]_{PB} = -2x_1 (\partial\phi / \partial x_1);$$

$$[\Gamma_7, \Gamma_9]_{PB} = -2x_2\,(\partial\phi/\partial x_2); [\Gamma_7, \Gamma_{10}]_{PB} = -2x_3\,(\partial\phi/\partial x_3).$$

Further, a differential equation like (3.77), for the TD function $\lambda_7\,(t)$, can now be derived as

$$\dot{\lambda}_7\phi - (\beta\psi - \lambda_7)\,[\,p_1(\partial\phi/\partial x_1) + p_2(\partial\phi/\partial x_2) + p_3(\partial\phi/\partial x_3)\,]$$

$$+\frac{1}{2}\beta\dot{\psi}\,[\,x_1(\partial\phi/\partial x_1) + x_2(\partial\phi/\partial x_2) + x_3(\partial\phi/\partial x_2)\,] = 0, \quad (6.19)$$

where we have chosen $\lambda_1 = \lambda_2 = \psi\,(t) = \lambda_3$ (say) and other λ_k's (except λ_7) are obtained as before. For λ_7, setting $\lambda_7 = \beta\psi$ reduces (6.19) to the form

$$\delta\,(t)\,\phi + x_1\,(\partial\phi/\partial x_1) + x_2\,(\partial\phi/\partial x_2) + x_3\,(\partial\phi/\partial x_3) = 0, \quad (6.20)$$

with $\delta\,(t) = 2 + 2\,\beta\psi\,/\,(\beta\dot{\psi})$ as before for the 2D case. Now, in addition to two constraining relations (3.76a,b) there appears a third one as $(\dddot{\psi}/4) + 2\alpha_3\dot{\psi} + \dot{\alpha}_3\psi = 0$. For the two separable (under addition and multiplication) forms of the solution ϕ of (6.20), namely

$$\phi\,(x_1, x_2, x_3) = k_1\,x_1^{-\delta} + k_2\,x_2^{-\delta} + k_3\,x_3^{-\delta}, \quad (6.21)$$

and

$$\phi\,(x_1, x_2, x_3) = k_4\,x_1^{-c-\delta}\,x_2^{c-c'}\,x_3^{c'} + k_5\,x_2^{-c-\delta}\,x_3^{c-c'}\,x_1^{c'} + k_6\,x_3^{-c-\delta}x_1^{c-c'}x_2^{c'}, \quad (6.22)$$

the corresponding second-order invariants now turn out to be

$$I = \frac{1}{2}\psi\,(p_1^2 + p_2^2 + p_3^2) + (\frac{1}{4}\ddot{\psi} + \alpha_1\psi)\,x_1^2 + (\frac{1}{4}\ddot{\psi} + \alpha_2\psi)\,x_2^2 + (\frac{1}{4}\ddot{\psi} +$$

$$\alpha_3\psi)\,x_3^2 + \beta\psi\,(k_1\,x_1^{-\delta} + k_2\,x_2^{-\delta} + k_3\,x_3^{-\delta}) - \frac{1}{2}\dot{\psi}\,(x_1p_1 + x_2p_2 + x_3p_3) \quad (6.23)$$

and

$$I = \frac{1}{2}\psi\,(p_1^2 + p_2^2 + p_3^2) + (\frac{1}{4}\ddot{\psi} + \alpha_1\psi)\,x_1^2 + (\frac{1}{4}\ddot{\psi} + \alpha_2\psi)\,x_2^2$$

$$+(\frac{1}{4}\ddot{\psi} + \alpha_3\psi)\,x_3^2 + \beta\psi\left[k_4\left(\frac{x_2}{x_1}\right)^c\left(\frac{x_3}{x_1}\right)^{c'} x_1^{-\delta} + k_5\left(\frac{x_3}{x_2}\right)^c\left(\frac{x_1}{x_2}\right)^{c'} x_2^{-\delta}\right.$$

$$+ k_6 \left(\frac{x_1}{x_3}\right)^c \left(\frac{x_2}{x_3}\right)^{c'} x_3^{-\delta}\Bigg] - \frac{1}{2}\,\dot{\psi}\,(x_1\, p_1 + x_2\, p_2 + x_3\, p_3), \tag{6.24}$$

respectively. Here c and c' are the separation and k_i ($i=1, .., 6$) are the integration constants.

The important point to be noted here is about the nature of the rationale regarding the powers of the coupling term in (6.18) for the case (6.22). In fact, if we write the solution (6.22) as $\phi = k\, x_1^m\, x_2^n\, x_3^l$, then for the case when β (and hence δ) is TD, the first invariant is found to exist only for $m + n + l = -\delta$, a situation similar to that for the 2D case. However, if β is TID, then $\delta = 2$ and we have $m + n + l = -2$. Again for $m = n = l = 1$, one arrives at an interesting case of Inozemtsev-type coupling [33], viz., $\phi = k\,(x_1\, x_2\, x_3)^{-2/3}$, but with harmonic terms in the 3D case. Also, such a system now admits second-order invariant. A study of the system (6.18) may also involve further generalizations of the Ermakov systems (cf. Sect. 3.5.2). We avoid such details here and leave as an exercise to the interested reader.

6.1.3 Time- and Momentum-Dependent Systems in 3D

A 3D analogue of the Hamiltonian system discussed in Sect. 3.3.2 (c) in the form

$$H = \frac{1}{2}\,(p_1^2 + p_2^2 + p_3^2) + \alpha_1\,(t)\, x_1^2 + \alpha_2\,(t)\, x_2^2 + \alpha_3\,(t)\, x_3^2$$

$$+ \beta\,(t)\, g\,(p_1, p_2, p_3),$$

$$\equiv \Gamma_1 + \Gamma_2 + \Gamma_3 + \alpha_1\,(t)\,\Gamma_4 + \alpha_2\,(t)\,\Gamma_5 + \alpha_3\,(t)\,\Gamma_6 + \beta\,(t)\,\Gamma_7 ,$$

can also be analysed here. Note that the additional Γ_i's needed to close the algebra turn out to be $\Gamma_8 = -2p_1x_1, \Gamma_9 = -2p_2x_2$, and $\Gamma_{10} = -2p_3x_3$, with corresponding h_i's as $h_8 = h_9 = h_{10} = 0$. The nonvanishing Poisson brackets can be computed as before and finally the differential equation satisfied by λ_7, now becomes (see KPGM [85])

$$\dot{\lambda}_7\, g = 2\,[\,(\lambda_7\alpha_1 - \beta\lambda_4)\, x_1\,(\partial g / \partial p_1) + (\lambda_7\alpha_2 - \beta\lambda_5)\, x_2\,(\partial g / \partial p_2)$$

$$+ (\lambda_7\alpha_3 - \beta\lambda_6)\, x_3\,(\partial g / \partial p_3)\,] + 2\,\beta\,[\lambda_8\, p_1\,(\partial g / \partial p_1)$$

$$+ \lambda_9\, p_2\,(\partial g / \partial p_2) + \lambda_{10}\, p_3\,(\partial g / \partial p_3)\,] .$$

For $\alpha_1 = \alpha_2 = \alpha_3 = \alpha$ (say) and $\lambda_7 = \beta\psi + (\beta\,\ddot{\psi}/4\alpha)$, the above equation reduces (see KPGM [85]) to the form $\Delta(t)\; g = p_1\,(\partial g/\partial p_1) + p_2\,(\partial g/\partial p_2) + p_3\,(\partial g/\partial p_3)$ where $\Delta(t) = 2 + 2\,(\dot{\beta}\psi/\beta\dot{\psi}) + (\dddot{\psi}/2\alpha\dot{\psi}) + \ddot{\psi}\,(\dot{\beta}\alpha - \dot{\alpha}\beta)/(2\beta\,\alpha^2\dot{\psi})$, as for the 2D case. The ansatz for the separable solutions of this PDE in the forms $g\,(p_1, p_2, p_3) = k_1 p_1^{\Delta} + k_2 p_2^{\Delta} + k_3\, p_3^{\Delta}$ and $g\,(p_1, p_2, p_3) = k_4 p_1^{-c-c'+\Delta}\, p_2^{c}\, p_3^{c'}$ give rise to the invariants

$$I = \frac{1}{2}\,\psi\,(p_1^2 + p_2^2 + p_3^2) + (\frac{1}{4}\,\ddot{\psi} + \alpha\psi)\,(x_1^2 + x_2^2 + x_3^2) + (\frac{\beta}{4\alpha}\,\ddot{\psi} + \beta\psi).$$

$$(k_1 p_1^{\Delta} + k_2 p_2^{\Delta} + k_3 p_3^{\Delta}) - \frac{1}{2}\,\dot{\psi}\,(p_1 x_1 + p_2 x_2 + p_3 x_3),$$

and

$$I = \frac{1}{2}\,\psi\left(p_1^2 + p_2^2 + p_3^2\right) + \left(\frac{1}{4}\,\ddot{\psi} + \alpha\psi\right)\left(x_1^2 + x_2^2 + x_3^2\right) + \left(\frac{\beta}{4\alpha}\,\ddot{\psi} + \beta\psi\right).$$

$$(k_4 p_1^{-c-c'+\Delta}\, p_2^{c}\, p_3^{c'} + k_5\, p_1^{-c'}\, p_2^{-c-c'+\Delta}\, p_3^{c'} + k_6\, p_1^{c}\, p_2^{c'}\, p_3^{-c-c'+\Delta})$$

$$-\frac{1}{2}\,\dot{\psi}\,(p_1 x_1 + p_2 x_2 + p_3 x_3),$$

respectively. Here ψ and α satisfy the constraining relation

$$\dddot{\psi} + 8\alpha\,\dot{\psi} + 4\dot{\alpha}\,\psi = 0\,.$$

It may be of interest to compare the present results for the momentum-dependent case with the coordinate-dependent ones of Sect. 6.1.2. Note that in the latter case the invariant exists for the form of the coupling term as $\bar{\phi} = \beta(t)\, x_1^m\, x_2^m\, x_3^l$, with $m + n + l = -\,\delta$, whereas in the former case it exists for $m + n + l = \Delta$ in $\bar{g} = \beta(t)\, p_1^m\, p_2^n\, p_3^l$. Here m, n, l are some real numbers, not necessarily the integers. Further, for β as TID, the δ reduces to a fixed number 2. On the other hand, Δ(t) still retains time-dependence in the form $\Delta\,(t) = 2 + (\alpha\,\dddot{\psi} - \dot{\alpha}\,\ddot{\psi})/(2\,\dot{\psi}\,\alpha^2)$. As a matter of fact, for the momentum-dependent case $\Delta\,(t)$ also depends on the TD frequency $\alpha\,(t)$, although the same is not the case with the coordinate-dependent coupling. Thus, for the existence of the invariant the rationale for the choices of coordinate-and momentum-dependent couplings in the Hamiltonian are quite different. One can as well investigate the mixed (both coordinate-and momentum-dependent) couplings in the Hamiltonian within this framework (cf. Sect. 7.2.6 (*d*) for the 2D case).

6.2 QUANTUM ASPECT

6.2.1 TID Systems in 3D

Here, we first discuss the method of separation of variables in polar coordinates for NC potentials in 3D and then make a few remarks on other available methods.

Khare and Bhaduri [131], as mentioned in Sect. 4.4.1, have studied the solution of the TID SE

$$\left[-\left(\frac{\partial^2\Psi}{\partial r^2}+\frac{2}{r}\frac{\partial\Psi}{\partial r}\right)-\frac{1}{r^2}\left(\frac{\partial^2\Psi}{\partial\theta^2}+\cot\theta\frac{\partial\Psi}{\partial\theta}\right)-\frac{1}{r^2\sin^2\theta}\frac{\partial^2\Psi}{\partial\varphi^2}\right]=(E-V)\Psi, \tag{6.25}$$

for a class of 3D potentials expressible in the form,

$$V(r,\theta,\varphi)=U(r)+(U_1(\theta)/r^2)+U_2(\varphi)/(r^2\sin^2\theta). \tag{6.26}$$

By writing $\Psi(r,\theta,\varphi)$ as

$$\Psi(r,\theta,\varphi)=\frac{u(r)}{r}\cdot\frac{H(\theta)}{(\sin\theta)^{1/2}}\cdot K(\varphi), \tag{6.27}$$

eq. (6.25) can be separated into three ODE's obeyed by the functions $u(r)$, $H(\theta)$ and $K(\varphi)$ which, respectively correspond to $r-$, $\theta-$ and $\varphi-$variables. These three equations can then be solved analytically for suitable choices of the potentials. A typical case investigated by Khare and Bhaduri[131] corresponds to the potential (6.26) with

$$U(r)=\frac{1}{4}\omega^2r^2+\delta/r^2;\; U_1(\theta)=C\,\mathrm{cosec}^2\theta+D\sec^2\theta,$$

$$U_2(\varphi)=G\,\mathrm{cosec}^2 p\varphi+F\sec^2 p\varphi, \tag{6.28}$$

where p is an integer. For these forms of $U(r), U_1(\theta), U_2(\varphi)$ and for various values of p, the potential (6.26) attains several standard forms used in the literature. Finally, the analytic forms of the solution of the $K-$, $H-$ and $u-$ equations for this potential can respectively be written as

$$K_{n_1}(\varphi)=(\sin p\varphi)^{\alpha}(\cos p\varphi)^{\beta}\,P_{n_1}^{\alpha-1/2,\beta-1/2}(\cos 2p\varphi),$$

$$H_{n_2}(\theta)=(\sin\theta)^{\bar{\alpha}}(\cos\theta)^{\bar{\beta}}\,P_{n_2}^{\bar{\alpha}-1/2,\bar{\beta}-1/2}(\sin 2\theta),$$

$$u_{n,n_1,n_2}(r) = r^{((\delta+\ell^2)^{1/2}+(1/2))} . \exp\left(-\frac{1}{4}\omega r^2\right) . L_n^{(\delta+\ell^2)^{1/2}}\left(\frac{1}{2}\omega r^2\right),$$

with the corresponding eigenvalues given by

$$E_{n,n_1,n_2} = [\,(2n+1) + (\delta+\ell^2)^{1/2}\,]\,\omega, \tag{6.29}$$

where

$$\ell^2 = \left[\ (2n_2+1) + \sqrt{D+(1/4)} + \left\{C + \left(\sqrt{F+(p^2/4)} + \sqrt{G+(p^2/4)} + p\,(2n_1+1)\right)^2\right\}^{1/2}\right]^2. \tag{6.30}$$

Here, $P_n^{m,l}(x)$ *and* $L_n^m(x)$ are the Jacobi and Laguerre polynomials, respectively. Note that eq. (6.30) is some sort of constraining relation among the potential parameters.

The study of dynamical symmetries of the SE can also be useful to the 3D case. In fact, several systems allowing three or more quadratic constants of motion (in the same spirit as discussed for the 2D case in Sect. 4.7) are found by Makarov et al [147]. Following are some potentials for which the SE separates in the spherical coordinate system and at least in one additional system of coordinates

$$V_1(x_1, x_2, x_3) = \alpha\,(x_1^2 + x_2^2 + x_3^2) + \frac{\beta_1}{x_1^2} + \frac{\beta_2}{x_2^2} + \frac{\beta_3}{x_3^2},$$

(rectangular, elliptic cylinderical, ellipsoidal),

$$V_2(x_1, x_2, x_3) = \frac{\alpha}{x_2^2} + \frac{\beta}{x_3^2} + \frac{\gamma\,x_1}{x_2^2\,(x_1^2 + x_2^2)^{1/2}}, \quad \text{(parabolic cylinderical)}$$

$$V_3(x_1, x_2, x_3) = \alpha\,(x_1^2 + x_2^2 + x_3^2) + \frac{\beta}{x_3^2} + \frac{h\,(x_2/x_1)}{x_1^2 + x_2^2},$$

(circular cylinderical, both prolate and oblate spheroidal)

$$V_4(x_1, x_2, x_3) = \frac{\alpha}{r} + \frac{\beta\cos\theta}{r^2\sin^2\theta} + \frac{h\,(\tan\varphi)}{r^2\sin^2\theta}, \quad \text{(parabolic cylinderical)}$$

$$V_5(x_1, x_2, x_3) = f(r) + \frac{\beta_1}{x_1^2} + \frac{\beta_2}{x_2^2} + \frac{\beta_3}{x_3^2}, \quad \text{(conical).}$$

Interestingly, for all these potentials the SE admits closed form solutions alongwith complete spectrum of eigenvalues. In this context, however, the question remains as to how to get a complete set of involuting invariants (if they exist) towards the integrability of the higher-dimensional systems.

Finally, it may be mentioned that some approximation methods (such as Hill-determinant or continued fraction method [109]) for a certain class of potentials can also give rise to closed-form solutions. In this context we mention the work of Witwit [142a] for the 3D systems and of Znojil [102] for 1D systems, outlined earlier in Chap. 4.

6.2.2 TD Systems in 3D

A detailed study of the quantum aspect of TD systems in 3D has not been carried out in the literature to the best of my knowledge. In this context, no doubt the simple method pursued by Ray [80] and by Kaushal [166] in 1D and 2D, respectively (cf. Sect. 5.2.3 and 5.4), can easily be extended to the 3D case; but only a limited class of TD potentials can be explored within this framework. On the other hand, the involved approaches like that of Boyer [169], Boyer et al [169] or of Truax [160], if extended to higher dimensions, may turn out to be more elegant in this respect.

6.3 CONCLUDING DISCUSSION

In the preceding sections we have rather demonstrated the extension of some of the methods used for the 2D systems, to the case of 3D systems. This has been possible more for the classical case than for the quantum case. In fact, in the quantum context such methods are yet to be developed.

In particular, the extension of the rationalization method, the methods of Lie symmetries and Lie algebraic approach are discussed for 3D systems. Several other methods presented in Chaps. 2 and 3, can also be extended to the 3D case in principle; but in practice the magnitude of complexity in studying these systems increases with dimensions. Such difficulties become particularly more prominent when we look for the higher-order invariants in the classical case and their role in the quantum case. Similar is the case with the study of quantum aspect of these systems.

Finally, it may be mentioned that the study of 3D or higher-dimensional systems should not always be considered as an extension of that of 2D systems at least at the classical level. The reason being the following: As it can be noticed [16,141] that for 1D systems the energy shell and tori are the same one-dimensional manifold and for 2D systems, the two-dimensional tori are embedded in the three-dimensional energy shell which means they divide the energy shell into inside and outside. (Note that for an N-dimensional integrable system, the phase space is $2N$-dimensional and the dimensionality of energy shell is $(2N-1)$ and that of the tori is N.) As a result, if somehow a "gap" between tori exists (which occurs for nonintegrable systems), then a trajectory in that gap cannot escape from it. However, for the three- or higher-dimensional systems the situation is different because of possible availability of invariants other than the Hamiltonian. The trajectories in "gaps" between higher-dimensional tori can escape to other regions of the energy shell. This gives rise to the phenomenon of 'Arnold diffusion'. (For the details, see Ref. [141] and the references therein). Thus, at least for nonintegrable systems in two- and higher- dimensions separate studies are required. Moreover, in this paragraph the space dimensions of the system should not be confused with the dimensionality of those of energy shell or of tori.

7

Role and Scope of Dynamical Invariants in Physical Problems: Interpretation and Applications

It is true that the invariants when defined in a broader sense play an important role in different branches of mathematics and mathematical physics, but somehow the description of physical reality limits their applications in physics and other allied sciences. In this respect while the role and the scope of the dynamical invariants for the 1D systems has been discussed at great length in the literature, it is not explored to that extent for the two and higher dimensional systems. In this chapter, we briefly highlight some possible physical interpretations known for the invariant for the TD HO system in 1D and also discuss some physical situations in different branches of mathematical sciences where not only the role of these invariants becomes transparent but also their studies may find some applications. Some possible interpretations of dynamical invariants are outlined in the next Section. In Sect. 7.2, various applications of the knowledge of these invariants are discussed. In Sect. 7.3, the role of these invariants is briefly discussed in the context of more recent topic of phases of the quantum wavefunction.

7.1 PHYSICAL INTERPRETATIONS OF DYNAMICAL INVARIANTS

From the point of view of having an in-depth study of a dynamical system, no doubt it is desirable to know all of its permissible invariants, if they exist, but as far as the assigning of the physical meaning to these invariants is concerned it has not been possible even to all the available ones. Further, the assignment of such a physical meaning to a complicated mathematical form of an invariant although appears to be difficult, in general, but for the polynomial form such possibilities have been explored. In particular, we list here some plausible interpretations suggested for the form (3.5) which corresponds to a TD HO system.

(i) According to Eliezer and Gray [170], the constancy of I is equivalent to the constancy of the angular momentum associated with the auxiliary eq. (3.6). The motion in a straight line described by $\ddot{x} + \omega^2(t)\, x = 0$, can be viewed as the projection of 2D motion of a particle under an attractive centre of force. Then the auxiliary motion is described by

$$\ddot{\vec{r}} + \omega^2(t)\, \vec{r} = 0, \tag{7.1}$$

where the vector $\vec{r}$ has Cartesian components (x,y). Solving eq. (7.1) radially and transversely using polar coordinates (ρ,θ), where $\rho = |\vec{r}|$, $x = \rho\cos\theta$, $y = \rho\sin\theta$, it gives

$$\ddot{\rho} - \rho\,\dot{\theta}^2 + \omega^2(t)\,\rho = 0\,;\ (1/\rho)\,(d(\rho^2\dot{\theta})/dt) = 0. \qquad (7.2\text{ a, b})$$

The latter equation implies $\rho^2\dot{\theta} = \ell$, where ℓ is the angular momentum constant. Now, using this result in (7.2a), one obtains the same equation as (3.6) with k replaced by ℓ^2 and after using $x = \rho\cos\theta$, $p = \dot{x}$, the invariant I of eq. (3.5) reduces to $I = (1/2)\,\ell^2$. This shows that the constancy of I is equivalent to the constancy of the angular momentum of the auxiliary motion.

(ii) Takayama [78] has discussed the physical meaning of the invariant (3.5) for a real system—the forced betatron oscillation seen in the accelerator and storage rings. In this case the motion is described by an equation of the type (3.10), where $g(t)$ is the external TD force. The conserved quantity, obtained for this system in the form of "affine" invariant namely, $\dot{x}_1\, x_2 - x_1\, \dot{x}_2 =$ const., where $x_1(t)$ and $x_2(t)$ are arbitrary solutions of eq. (3.19), does not solve any purpose. On the other hand, a physical interpretation is looked for a form similar to (3.5) using the concept of the equilibrium orbit well known in accelerator physics. In fact, the equilibrium orbit (u,v) is the particular solution of the canonical equations

$$(du/dt) = (\partial H/\partial v) = v\,; \tag{7.3}$$

$$(dv/dt) = -\,(\partial H/\partial u) = -\,\omega^2(t)\,u + g\,(t), \tag{7.4}$$

where the Hamiltonian, H, has the form

$$H = (1/2)\,[\,p^2 + \omega^2(t)\,x^2\,] - g(t)\,x. \tag{7.5}$$

As a matter of fact a linear transformation, $x = u + X, p = v + P$, converts the Hamiltonian (7.5) into the form (3.4) and the corresponding invariant into the form (3.5), for which the physical interpretation becomes easier.

(iii) Kaushal and Korsch [66] interpreted the invariant I as a mapping between the two similar dynamical systems satisfying a definite force law. It is shown that the invariant I can be written in the form

$$I = (1/2)\,\{\bar{k}\,(x/\bar{x})^2 + k\,(\bar{x}/x)^2 + (\bar{x}\,\dot{x} - \dot{\bar{x}}\,x)^2\}, \tag{7.6}$$

for the Hamiltonian

$$H = (1/2)\,[\,p^2 + \omega^2(t)\,x^2 + \;k / x^2] \tag{7.7}$$

with $\bar{x}$ as the solution of the auxiliary equation

$$\ddot{\bar{x}} + \omega^2(t)\,\bar{x} = \bar{k} / \bar{x}^3. \tag{7.8}$$

Note that eq. (7.8) is actually the equation of motion corresponding to a similar Hamiltonian system

$$\bar{H} = (1/2)\,[\,\bar{p}^2 + \omega^2(t)\,\bar{x}^2 + \bar{k} / \bar{x}^2\,], \tag{7.9}$$

where k in (7.7) and $\bar{k}$ in (7.9) are constants. Thus, the equation of motion corresponding to $\bar{H}$ is the same as the auxiliary equation for H or vice versa. Here, I shows the correspondence between the Hamiltonians H and $\bar{H}$. In fact, the invariant (7.6) is a common constant of motion w.r.t. H and $\bar{H}$, so that we have

$$[I, H]_{\mathrm{PB}(x,p)} = [\,I, \bar{H}\,]_{\mathrm{PB}(\bar{x},\bar{p})}, \tag{7.10}$$

i.e. I only generates a mapping between H and $\bar{H}$. It may be mentioned that in the spirit of the results of Sect. 3.3.2, (cf. eq. (3.81)), the time-dependence of k and $\bar{k}$ can also be investigated in such an interpretation of I, of course for the corresponding 1D system.

(iv) An alternative interpretation of the invariant (7.6) can also be found by realising that it has a Hamiltonian structure [171] in terms of the newly defined coordinate η and the time parameter τ. Note that I in (7.6) can be expressed in an alternative form as

$$I = (1/2)\,[\,k\,(x/\bar{x})^2 + \bar{k}\,(\bar{x}/x)^2 + \bar{x}^4\,\{(d/dt)\,(x/\bar{x})\}^2]. \tag{7.11}$$

Now, after defining $\eta = (x/\bar{x}), \sqrt{2}\,d\tau = dt/\bar{x}^2$, $2I$ from (7.11) takes the form

$$2I = k\,\eta^2 + \bar{k} / \eta^2 + (1/2)\,(d\eta / d\tau)^2. \tag{7.12}$$

It is interesting to note that $2I$ here has a Hamiltonian structure with the potential term very similar to that of H or $\overline{H}$ (cf. eqs. (7.7) and (7.9)) and the kinetic term as $(1/2)\,(d\eta/d\tau)^2$.

7.2 APPLICATIONS OF DYNAMICAL INVARIANTS

In this Section we highlight the applications of the knowledge of dynamical invariants (particularly that of the one derived for the TD HO system in 1D) in the context of quantum mechanics, Feynman propagator (using path integral technique), cosmology, relativistic TD Hamiltonian systems, obtaining the solution of a certain class of nonlinear differential equations, and of several other fields such as biophysics, plasma physics and field theories etc.

7.2.1 Quantum Mechanics

The solution of the classical TD HO problem has suggested an alternative method for developing a Schrodinger-type quantum mechanics and a WKB-type semiclassical quantization condition. In particular, Korsch and his co-workers [172, 173], and Lee [174] have used these methods to obtain the solution of some of the physical problems in an alternative manner. Some of these aspects are discussed here briefly:

(*a*) *Milne's Equation and WKB-type quantization*

It can be seen that just a simple replacement of ρ, x, t and $\omega(t)$ in the results for the TD HO system (cf. eqs. (3.4) – (3.6) and (3.19)) by w, u, x, and $k(x)$, respectively leads to following equations:

$$w''(x) + k^2(x)\, w(x) = 1/w^3(x), \tag{7.13}$$

$$u''(x) + k^2(x)\, u(x) = 0. \tag{7.14}$$

If we use $k(x) = [2m\,(E - V(x))/\hbar^2]^{1/2}$, then eq. (7.14) yields the form of the well known Schrodinger equation and eq. (7.13) the form of which is known as Milne's equation in the literature [172, 174]. Now, if we know a particular solution of (7.13), the general solution of (7.14) can be written as [172]

$$u(x) = c\, w(x) \sin\left(\int^x w^{-2}(x')\, dx' - b\right), \tag{7.15}$$

where c and b are arbitrary constants. Alternatively, the general solution of (7.13) can be obtained in terms of linearly independent solutions u_1 and u_2 of (7.14) as [72, 170]

$$w(x) = [A\, u_1^2(x) + B\, u_2^2(x) + 2C\, u_1(x)\, u_2(x)]^{1/2}, \tag{7.16}$$

where A, B and C are constants related to the Wronskian W of u_1 and u_2 by $AB - C^2 = W^{-2}$.

Using the boundary condition on the wave function for the bound state to exist, one can arrive at

$$\int_{-\infty}^{\infty} w^{-2}(x)\,dx = (n+1)\,\pi, \quad (n = 0, 1, 2, ..), \tag{7.17}$$

which is termed [172, 174] as the *Milne's quantization condition* for the energy levels E_n. In the semiclassical limit ($\hbar \to 0$) of Milne's equation (i.e. after neglecting the w''-terms in (7.13) one obtains $w(x) \approx [k(x)]^{-1/2}$, a result valid in the classically allowed region and breaks down at the classical turning points. In this case the condition [7.17] takes the form

$$\int_{-\infty}^{\infty} k(x)\,dx = (n+\frac{1}{2})\,\pi, \quad (n = 0, 1, 2, ..), \tag{7.18}$$

where the missing $(\pi/2)$-term accounts for the contribution from the classically forbidden regions. Several applications of this new quantization rule and its possible generalization to the complex energy case are discussed by Korsch and Laurent [172] and Korsch et al. [173].

(b) Quantum mechanics as a multidimensional Ermakov theory

Lee [174] offers a new dimension to the applications of the Milne's quantization condition and the Ermakov theory mentioned above. In particular, the close connection between the classical Ermakov theory and the Milne's quantization condition is exploited to the extent of finding a common mathematical framework within which an explanation of both classical particle mechanics and quantum wave mechanics can be sought. As a result, a physical interpretation of the Ermakov invariant in terms of the theory of wave-particle duality is suggested by Lee [174]. While it may be desirable to pursue several other problems within this framework, the problems pertaining to quantum ray optics and hydrogen atom have already been looked into by Lee. As a matter of fact in the multidimensional Ermakov theory, there seems to exist [171] a common basis for both the Schrodinger equation in quantum mechanics and the Riccati-type equation satisfied by the superpotential in the supersymmetric quantum mechanics.

7.2.2 Feynman Propagator (Using Path Integral Technique)

The existence of an invariant for a dynamical system also simplifies the calculation of the Feynman propagator. In fact the Feynman propagator, while already containing the spirit of the quantum superposition principle via integration over various paths, provides an alternative route from classical to quantum description of a system. Lawande et al [159, 175, 19] have studied in detail the role played by the invariants in the propagator theory using a large class of potentials. It is noticed that a great simplification, in this case, arises if the invariant is assumed to be of second order in momenta. Further, explicit path-integral calculations have shown that the propagators in general admit expansions in terms of the eigenfunctions of the invariant operator. This, in fact, allows, as mentioned in Sect. 5.3, the Feynman propagator to be expressed in terms of the eigenfunctions of the Ermakov invariant in an exact manner.

A variety of dynamical systems have been studied by Lawande and his coworkers [175, 176] by extending the propagator theory to several dimensions of applications. For the details we refer to the excellent review by Khandekar and Lawande [19]. Here, we just demonstrate the use of the Ermakov invariant in the calculation of the propagator for a simple system described by the Lagrangian

$$L = (1/2)\,\dot{x}^2 - (\ddot{\rho}\,\alpha/\rho - \ddot{\alpha})\,x + (\ddot{\rho}/2\rho)\,x^2 - (1/\rho^2)\,F((x-\alpha)/\rho), \tag{7.19}$$

and which possesses a second order invariant

$$I(x,p,t) = (1/2)\,[\,\rho(p-\dot{\alpha}) - \dot{\rho}(x-\alpha)\,]^2 + F((x-\alpha)/\rho). \tag{7.20}$$

Here $\rho(t)$, $\alpha(t)$ and $F((x-\alpha)/\rho)$ are arbitrary functions of their arguments. It is possible to write (7.19) in the form,

$$L = (d\chi/dt) + L_0, \tag{7.21}$$

where L_o is a new Lagrangian given by

$$L_o = (1/2)\,\rho^2\,[\,(d/dt)((x-\alpha)/\rho)\,]^2 - (1/\rho^2)\,F((x-\alpha)/\rho),$$

and

$$\chi = (\dot{\rho}/2\rho)\,x^2 + ((\dot{\alpha}\rho - \alpha\dot{\rho})/\rho)\,x - (1/2)\int \rho^2\,[(d/dt)(\alpha/\rho)]^2 dt.$$

The Feynman propagator, $K(x'', t''; x', t')$, defined as the quantum mechanical amplitude for finding a particle at the position x'' at the time t'' if the particle had been at x' at an earlier time t', is expressed by (cf. Sect. 5.3)

$$K(x'', t''; x', t') = \int \exp\{(i/\hbar)\int_{t'}^{t''} L\,dt\}\,\mathcal{D}x(t), \tag{5.33}$$

where $\mathcal{D}x(t)$ is the usual Feynman differential measure. After carrying out some lengthy calculations, finally, the propagator K can be expressed as [176]

$$K(x'',t''; x', t') = (\rho'\,\rho'')^{-1/2}\exp[\,(i/\hbar)\,\{\chi(t'') - \chi(t')\}\,]\,\overline{K}_o(\xi'', \tau''; \xi', \tau') \tag{7.22}$$

where

$$\overline{K}_o(\xi'', \tau''; \xi', \tau') = \int \exp((i/\hbar)\int_{\tau'}^{\tau''} \overline{L}_o\,d\tau)\,\mathcal{D}\xi(t),$$

with

$$\bar{L}_0 = \bar{L}_0(\xi, d\xi / d\tau) = (1/2)(d\xi / d\tau)^2 - F(\xi), \ (\xi = x / \rho).$$

Thus, it becomes clear that the propagator for TD system is related to the propagator for an associated TID system corresponding to the Lagrangian $\bar{L}_0$ in the new space-time variables (ξ, τ). Further, note that the invariant [7.20] is basically the Hamiltonian H_0 associated with the Lagrangian $\bar{L}_0$. Khandekar and Lawande [176] have also obtained the Feynman propagator in an exact and closed form for NC potentials.

7.2.3 Cosmological Applications

The importance of dynamical invariants in the fields of cosmology and astrophysics is well known [54, 177]. In particular, Berger [178], Misner [179], and Ray [180a] have studied in detail the relationship between the particle number present in a cosmological model and an adiabatic invariant. The knowledge of invariants, in fact, offers an alternative for calculating the particle production in cosmological models.

Here we discuss rather briefly the related adiabatic invariant which is defined as a semiclassical particle number, and the alternative method for calculating the particle production in models with an initial singularity. Once the particle number for each mode of the field is defined as an adiabatic invariant then it becomes interesting to relate this quantity to the parameters of the field mode near the initial singularity. Following Berger [178], the amplitude $\phi_{\vec{k}}$ for each mode $\vec{k}$ of a minimally coupled scalar field of mass m, satisfies

$$\ddot{\phi}_{\vec{k}} + \omega^2_{\vec{k}}(\tau)\phi_{\vec{k}} = 0, \tag{7.23}$$

where τ is a new time coordinate. In a regime in which the expansion or contraction time-scale of the universe greatly exceeds the period of the mode $\vec{k}$, the solution to eq. (7.23) can be obtained [178] using the WKB approximation as

$$\phi_{WKB} \sim \omega^{1/2} A \cos\left(\int^{\tau} \omega \, d\tau' + \xi\right) + \omega^{-1/2} B \sin\left(\int^{\tau} \omega \, d\tau' + \xi\right), \tag{7.31}$$

where ξ is a constant phase and A, B are arbitrary constants. If the energy of the mode $\vec{k}$ is formally defined as

$$E(t) = (1/2)(\dot{\phi}^2 + \omega^2(t))\,\phi^2), \tag{7.25}$$

then in the WKB limit there exists [60] an adiabatic invariant N of the type $N = \dot{E}_{WKB} / \omega$, which in turn reduces to the form

$$N = (1/2)(A^2 + B^2). \tag{7.26}$$

On the other hand, Ray [180] makes use of the available form of the invariant for the system (7.25) as

$$I = (1/2)\,[\,(\phi/\rho)^2 + (\rho\dot{\phi} - \phi\dot{\rho})^2],\tag{7.27}$$

with ρ satisfying

$$\ddot{\rho} + \omega^2(t)\,\rho = \rho^{-3}\tag{7.28}$$

He, as before, identifies N of (7.26) with I in the adiabatic regime. In a model that possesses an initial singularity at time $t = t_s$ such that $\omega\,(t_s) = 0$, the invariant is expected to take the same form as (7.27) with ϕ and ρ replaced by ϕ_s and ρ_s corresponding to $t = t_s$, and the solutions of (7.23) and (7.28) for this case are expressed as [180a]

$$\phi = q_o + p_o\,(t - t_S)\,;\rho = (1/\alpha)\,[1 + \alpha^4\{t - t_S + (\beta/\alpha)\}^2\,]^{1/2},\tag{7.29}$$

where q_o, p_o, α and β are the integration constants. Further, use of these results yields the invariant (7.27) in the form

$$N = (1/2)\,[\,\alpha^2\,q_o^2 + (1/\alpha^2 + \beta^2)\,p_0^2 - 2\alpha\beta\,q_0 p_0].\tag{7.30}$$

Clearly, the modified definition of the constants A and B in (7.24) as $A = p_0/\alpha$, $B = \alpha q_0 - \beta p_0$, shows the equivalence of the forms (7.30) and (7.26). Thus, the Ermakov invariant provides an interesting alternative for calculating the particle production in cosmological models. Since it is an exact invariant, its leading term in the adiabatic series is the particle number.

Another aspect of the role of invariant in stellar structure studies has recently been highlighted [180b] in connection with the use of Tolman-Oppenheimer-Volkoff (TOV) theory for this purpose. In fact, in the spirit of the works of Korsch and his coworkers [172, 173], Lee [174], Kaushal and Parashar [171], the existence of a space-invariant in the TOV theory is established for certain forms of the equation of state. In particular, the role of such an invariant is expected to be crucial in studying the stability of stellar objects using the equations of state which involve a certain type of parameter-dependence.

7.2.4 Relativistic TD Hamiltonian Systems

The role of classical dynamical invariants is also found to be important in some special types of relativistic TD systems. Recently, there has been an attempt to obtain [181] the invariant for the system described by the Hamiltonian

$$H = (p^2 + 1)^{1/2} + V\,(x,t),\tag{7.31}$$

in the form $I\,(x, p, t) \equiv \Psi\,(V(x, t), p)$. The search for the integrable systems admitting such an invariant, following the method of Giacomini [182], is finally reduced to the solution of the "potential" equation of the form

$$(dV/dF_1) + f(V) + F_1/(F_1^2 + 1)^{1/2} = 0, \tag{7.32}$$

where both F_1 and f are the arbitrary functions of V and they are mutually related. The invariants corresponding to the three cases namely $f(V) = 1; f(V) = V$ and $f(V) = \exp(V)$, are obtained by Martin and Bouquet [181], respectively as

$$I = P + \mathcal{V} + (p^2 + 1)^{1/2}; I = V(x,t)\exp(p) + \int_0^P du.u \exp(u) / g(u), \tag{7.33}$$

and

$$I = [\exp(-V(x,t)) - p] \exp(-p^2 + 1)^{1/2} - \int_0^P du.u^2 \exp(-g(u)) / g(u),$$

where $V(x, t) = \mathcal{V}(x+t)$ and $g(u) = (u^2+1)^{1/2}$. As an example, the case of TD relativistic HO system in the form

$$H_R = (p^2 + 1)^{1/2} + x^2/2t^2, \tag{7.34}$$

has been analysed. Interestingly, such a system does not turn out to be analytically integrable, mainly due to the special nature of the kinetic term in H_R.

7.2.5 Solution of a Class of Nonlinear Differential Equations

In one way or the other the knowledge of invariants also helps in investigating the solutions of a particular class of nonlinear differential equations. Note that for the system (damped TD HO)

$$\ddot{x} + P(t)\dot{x} + Q(t)x = 0, \tag{7.35}$$

the invariant turns out to be

$$I = h^2 (x/\rho)^2 + (\dot{\rho}x - \rho\dot{x}) \exp\left(2\int_0^t P(t)\,dt\right), \tag{7.36}$$

with $\rho(t)$ satisfying

$$\ddot{\rho} + P(t)\dot{\rho} + Q(t)\rho = (h^2/\rho^3)\exp\left(-2\int_0^t P(t)\,dt\right). \tag{7.37}$$

Eliezer and Gray [170] put these simple results in the form of the following theorem:

Theorem: If $y_1(x)$ and $y_2(x)$ are the linearly independent solutions of

$$y'' + P(x)y' + Q(x)y = 0, \tag{7.35'}$$

then the general solution of

$$y'' + P(x)y' + Q(x)y = (h^2/y^3)\exp\left(-2\int P(x)\,dx\right) \tag{7.37'}$$

may be written as (cf. eq. (7.16))

$$y = (A\, y_1^2 + B\, y_2^2 + 2C\, y_1 y_2)^{1/2}. \tag{7.38}$$

Here, A, B and C are arbitrary constants and are related through $AB - C^2 = (h^2/W^2) \exp (\int P(x)\, dx)$, with $W = y_1 y'_2 - y'_1 y_2$, defined as the Wronskian for (7.35′). Clearly, in eq. (7.35′) and (7.37′) the dependent and independent variables are redefined and accordingly the dots are replaced by the primes in the definition of the derivatives.

Pinney [89] studied a nonlinear equation of the type

$$y'' + P(x)\, y' + Q\,(x)\, y = AB\, W^2 y^{-3}, \tag{7.39}$$

which again is the modified version of the auxiliary eq. (3.6) corresponding to a damped TD HO. The solution to (7.39) is found to be $y = (Ay_1^2 + By_2^2)^{1/2}$ Several generalized versions of the solution (7.38) to eq. (7.37′) have been investigated [26, 183-185] in the literature. For example, Reid [183] obtained the solution of the differential equation

$$y'' + P\,(x)\, y' + Q\,(x)\, y = AB\,(m-1)\,(y_1 y_2)^{m-2}\, W^2 y^{1-2m}, \tag{7.40}$$

in the form $y = (Ay_1^m + By_2^m)^{1/2}$ where m is a constant assumed to be real and nonzero. Thomas [184] has shown that the function $y = (y_1 y_2)^{k/2}$, satisfies the nonlinear equation

$$y'' + P(x)\, y' + kQ(x)\, y = (1-l)\, y'^2 y^{-1} - (1/4)\, kW^2\, y^{1-4l}, \tag{7.41}$$

where k is assumed to be real and nonzero and $kl = 1$. As a next step of generalization of eq. (7.37′), the function

$$y = (Ay_1^m + By_2^m + mCy_1^j y_2^n)^{k/m},$$

where $m = j+n$, is found to represent the solution of a nonlinear differential equation of more general type, namely

$$y'' + P(x)\, y' + kQ\,(x)\, y = (1-l)\, y'^2 y^{-1} + k\, U\, W^2\, y^{1-2ml}, \tag{7.42}$$

where $kl = 1$ as before, and

$$U = C\, y_1^{j-2} y_2^{n-2} [(m-j-1)n\, Ay_1^n + (m-n-1)j\, By_2^n - Cnj\, y_1^j y_2^n] + (m-1)\, AB\,(y_1 y_2)^{m-2}.$$

It is obvious that the earlier equations can be obtained as special cases of this last equation. For further details of such studies we refer to the literature [89, 183-185]. Interestingly, these classical dynamical studies have also suggested some clue for the solution of several quantum problems. In particular, Burt and Reid [183] have studied the solution of a nonlinear Klein-Gordon equation.

So far in this subsection we discussed the solution of the nonlinear equation of the type (7.42) in terms of the solutions of eq. (7.35′). But it may be reminded here that we have already investigated in detail in Sect. 3.5 another dimension of applications of the knowledge of invariants in relation to the coupled nonlinear differential equations in terms of Ermakov systems.

7.2.6 Other Applications

In this subsection we briefly highlight the role which dynamical invariants can possibly play (and, in fact, have already been playing in some localized domains) in the fields of biophysics, plasma physics, field theories, and several others. In any case, the present survey of TD systems can help further for a better understanding of the respective phenomenon in these areas.

(a) *Biophysics:* In neurology different models have been proposed for the nerve impulse propagation in terms of some nonlinear PDE's. The common difficulty with these models is that they turn [186, 187] out to be nonintegrable. In the following we briefly discuss a generalized but completely integrable model for the nerve impulse propagation proposed and pursued by Rajagopal [188, 189] in a series of papers.

The basic equation governing the nerve impulse propagation is of the form of a nonlinear diffusion equation which, after introducing a phenomenological expression for the ion current, can be written as

$$V_{xx} - V_t = F(V)\,, \tag{7.43}$$

where $V(x,t)$ is the voltage across the membrane and V_t is the measure of the displacement current per unit length passing through the membrane. Here and in the following the subscripts indicate the variables w.r.t. which the partial derivatives are taken. Somehow eq. (7.43) fails to reproduce the important feature of pulse recovery which is necessary for a repeated firing of the fibre. An account of such a recovery variable, R, yields (7.43) in the form as

$$V_{xx} - V_t = F(V) + R_t \tag{7.44}$$

with $R_t = \in (V + a - bR)$. Here $\in$ is proportional to the temperature factor and a and b are constants.

In his generalization of the model Rajagopal uses two first order differential equations in place of (7.44). For a space clamped axon, it is assumed that the rate of change of membrane potential (X_t) depends linearly on Z (the current stimulus applied through the electrode to the axon) and on Y (the intrinsic current) and nonlinearly on the membrane potential X, thereby implying

$$X_t = -\,a'\,[F(X) - Y - Z]. \tag{7.45}$$

Further, the rate of change of intrinsic current (Y_t) is chosen as $Y_t = b'\,[G(X) - Y]$. Here a' and b' are arbitrary constants. Although the experiments [186,

187] suggest a cubic nonlinearity for $F(X)$ and an exponential one for $G(X)$, but Rajagopal [189] assumes the same form for $F(X)$, $Z(t)$ and $G(X)$ as

$$F(X) = k_1 X - k_2 X^3\,; \quad Z(t) = k_3 X - k_4 X^3; \quad G(X) = \alpha X - \beta X^3; \quad (7.46)$$

where k_i $(i = 1, 2, 3, 4)$ are arbitrary constants and $\alpha = k_1 - k_3$, $\beta = k_2 - k_4$. After a straightforward derivation, the ODE to be handled finally for the case of travelling waves turns out to be

$$d^2X / d\xi^2 = \gamma X - \delta X^3, \tag{7.47}$$

where $\gamma = \alpha a' + b'$, $\delta = \beta a'$ and $\xi = x - ut$ is the moving space variable with u as the wave velocity. Although the solution of (7.47) has been obtained explicitly in terms of elliptic functions but an account of time-dependence of γ and δ (depending upon the choice of the model) in (7.47) will bring in the knowledge of dynamical invariants discussed in Sect. 3.1.2.

(*b*) *Plasma physics*: Whenever the time varying (electromagnetic or simply magnetic) field is involved in understanding (theoretically as well as experimentally) a physical phenomenon, the corresponding dynamical system turns out to be a TD one. Further, the invariants may exist and can be constructed for such a system in an exact or an approximate manner depending upon the nature of the time-dependence. Such systems have been known [57] for a long time now in the fields of plasma physics and so called magnetic surfaces. We cut-short such discussions here. Only recently, the representation of magnetic field with toroidal topology in terms of field-line invariants have been studied by Lewis [190] and Lewis and Abraham-Shrauner [191]. As a matter of fact they make use of the Boozer's representation [192] for a magnetic field with toroidal topology in the form

$$\vec{B}(\vec{r}) = (\vec{\nabla}\psi_\circ) \times (\vec{\nabla}\theta_\circ) + (\vec{\nabla}\phi_\circ) \times (\vec{\nabla}\chi_\circ), \tag{7.48}$$

where $\theta_\circ(\vec{r})$, $\phi_\circ(\vec{r})$, $\chi_\circ(\vec{r})$ and $\psi_\circ(\vec{r})$, respectively, are the poloidal angle, toroidal angle, poloidal flux and toroidal flux functions which characterize the nature of the magnetic field. Note that any $\vec{B}(\vec{r})$ given by (7.48) is divergence-free i.e. $\vec{\nabla}\cdot\vec{B} = 0$. The main advantage of using (7.48) is that it allows direct representations in Hamiltonian form of the differential equation for the magnetic field lines. Thus, the representation (7.48) turns out to be a useful one as far as the applications to tokamaks and stellarators are concerned. For the details we refer to these works [190, 191].

(*c*) *Field theories:* The classical analogue of quantum field theories in a suitable choice of the gauge also gives rise [193] to some sort of dynamical systems. Savvidy [193] studied the Yang-Mills system described by the Lagrangian in the covariant form as

$$L = -(1/4)\, F^{\alpha}_{\mu\upsilon},\, F^{\alpha}_{\mu\upsilon}, \tag{7.49}$$

in the SU(2) case by resorting to the gauge $A_o^{\alpha} = 0$, $A_i^{\alpha} = A_i^{\alpha}(t)$. Here symbols have their usual meanings. This has led to a classical dynamical system described by the coupled nonlinear differential equations of the form

$$\ddot{f}^{\alpha} + (1/2)\sum_b (f^b)^2 f^{\alpha} = 0. \tag{7.49'}$$

Further simplifications in this case are achieved [194] by imposing an additional condition $A_3^{\alpha} = 0$ on the potentials. We shall return to these details in Appendix 1.

Recently, the Abelian Higgs model described by the Lagrangian

$$L = -(1/4)\, F_{\mu\upsilon}\, F_{\mu\upsilon} + (1/2)\,(D_{\mu}\phi)^{*}(D^{\mu}\phi) + C_2|\phi|^2 - C_4\,|\phi|^4, \tag{7.50}$$

has also been studied by Kumar and Khare [195] in (2+1) dimensions. They make an ansatz for the gauge and Higgs fields as

$$A_o\,(\vec{x},t) = 0,\ A_1\,(\vec{x},t) = A_2\,(\vec{x},t) = h(t)/\sqrt{2},$$

$$\phi\,(\vec{x},t) = \exp\,(i\omega(x+y))\; q_2(t).$$

Further, by defining $h(t) = q_1(t) + \sqrt{2}\,\omega / e$, the system of nonlinear equations obtained by these authors is

$$\ddot{q}_1(t) = -e^2 q_1\, q_2^2\ ;\ \ddot{q}_2(t) = 2\,C_2 q_2 - 4\,C_4 q_2^3 - e^2 q_1^2 q_2.$$

Clearly, such a system of equations can offer the example of a Ermakov type system under certain conditions.

(*d*) *Quantum optics and squeezed states:* It is well-known [196-199] that the HO with TD frequency provides a useful model for the description not only of coherent and squeezed states [197-198] but also of the coherent correlated states [199]. In this connection Malkin et al [198] introduced a TD complex invariant which is linear in momentum. Husimi [200] made a detailed study of TD quantal osillator with several of its variants. Man'ko [201] has emphasized that such studies of TDSE with reference to coherent states can be better carried out using the knowledge of this simple integral of motion.

In this case one looks for the solution of the SE ($\hbar = m = \omega\,(0) = 1$)

$$i\,\frac{\partial\Psi(x,t)}{\partial t} = -\frac{1}{2}\,\frac{\partial^2\Psi(x,t)}{\partial x^2} + \frac{1}{2}\,\omega^2(t)\,x^2\,\Psi(x,t). \tag{7.51}$$

The TD invariant (which is now an operator) considered [198] for this system is

$$A = \frac{i}{\sqrt{2}}\,[\epsilon\,(t)\,p - \dot{\epsilon}\,(t)\,x\,], \tag{7.52}$$

where the complex function $\in(t)$ satisfies the equation

$$\ddot{\in}(t)+\omega^2(t)\in(t)=0, \qquad (\in(0)=1, \dot{\in}(0)=i), \tag{7.53}$$

and the operator A conforms to $[A, A^\dagger]=1$. It is found [198, 201] that the packet solutions of the SE of Husumi [200] which now involve the function $\in(t)$ in the form

$$\Psi_\alpha(x,t)=\Psi_\circ(x,t).\exp\left\{-\frac{|\alpha|^2}{2}-\frac{\alpha^2\in^*(t)}{2\in(t)}+\frac{\sqrt{2}\,\alpha x}{\in(t)}\right\}, \tag{7.54}$$

where

$$\Psi_\circ(x,t)=\pi^{-1/4}\,[\in(t)]^{-1/2}.\exp[\,i\,\dot{\in}(t)\,x^2/2\in(t)\,],$$

may be introduced and interpreted as coherent states since they are eigenstates of the operator A given by (7.52). Here α is the complex number which appears in the definition of the coherent state $|\alpha>$ of Glauber [196].

The importance of the linear invariant (7.52) has also been noted by Man'ko [201] in the study of q-deformed Husimi packet solutions of the SE and accordingly a q-deformed integral of motion of the form

$$A_q=A\,[\sinh(\lambda A^\dagger A)/(A^\dagger A\sinh\lambda)]^{1/2},$$

where $\lambda=\ln q$, is also constructed. Here A_q satisfies $A_q A^\dagger_q - q A^\dagger_q A_q = q^{-A^\dagger A}$.

Now the question arises as what is the connection between the linear invariant (7.52) and the one (Ermakov invariant) studied in the classical case (cf. eq. (3.5)) or for that matter a linear structure studied by Takayama [78] and which also plays the role in solving the SE for TDHO potential (cf. Sect. 5.2.3). In fact while Lewis and Riesenfeld [64] first demonstrated that the same classical form of the invariant also works for the quantum case, Malkin et al [198] and Man'ko [201] have also looked into this problem. A special solution of eq. (7.53) (cited in Ref. [198]) as

$$\in(t) = |\in(t)|.\exp\{i\int^t d\tau/|\in(t)|^2\}, \tag{7.55}$$

converts eq. (7.53) into the form

$$\frac{d^2|\in(t)|}{dt^2}+\omega^2(t)\,|\in(t)|=1/|\in(t)|^3, \tag{7.56}$$

which, in fact, is the same as satisfied by the auxiliary function $\rho(t)$ (cf. eq. (3.6) for $k=1$). Also, the quantum analogue of the invariant I of eq. (3.5) can be considered as

$$I=A^\dagger(t)A(t)=\frac{1}{2}\{|\in(t)|^2p^2+|\dot{\in}(t)|^2x^2-\in^*\dot{\in}\,px-\in\dot{\in}^*\,xp\}, \tag{7.56}$$

connection, recently, the role of the quadratic invariant obtained for a TD harmonic plus inverse harmonic potential is highlighted by Kaushal and Parashar [201].

With a possibility to creating two-mode squeezed states, a 2D Hamiltonian in the form

$$H=\frac{1}{2}\sum_{k=1}^{2}\left\{\frac{p_k^2}{m(t)}+m(t)\,\omega_0^2\,x_k^2\right\}+\lambda(t)\,(x_1p_2-x_2p_1), \tag{7.57}$$

has also been investigated in the literature[202]. Here the time-dependence is considered in m and λ and not in ω_0 such that $m(t) = m_0 f(t)$, $\lambda(t) = \omega_0 g(t)$ with $f(0) = g(0) = 1$. While such a system is used to study the two-mode squeezing and photon generation phenomena, perhaps an account of another term either of the types (i) $\beta(t)\, x_1x_2$, (ii) $\beta(t)\, p_1p_2$, (iii) $\beta(t)\,(p_1x_2 + p_2x_1)$ in the Hamiltonian (7.57) can offer a better understanding of these phenomena from the point of view of correlations. Further, the knowledge of invariants constructed for some of these cases in Chap. 3, can also lead to some simplifications as far as the calculational details are concerned.

(e) Molecular and femto-chemistry: A number of TID systems (of both central and noncentral nature) can also arise [4] in the study of models in molecular chemistry. In this context, as mentioned before (cf. Chap. 4) not only the study of quantum aspects but also of classical aspects of such systems becomes important. It may be mentioned that the knowledge of dynamical invariants can play an important role in such studies.

In recent years, the study of ultra-fast dynamics of chemical bonds at the level of fermi-time scale has become [217] of great interest. One comes across a number of TD systems in two and higher dimensions in this case. In fact, these systems and related dynamical invariants, if exist, could be the theoretical manifestations of several experimentally tenable phenomena in femtochemistry. For details we refer to the recent publication by Zewail [217].

It may be mentioned that besides the above areas where the scope of invariants is highlighted, the role of dynamical invariants in the fields of condensed matter physics, statistical mechanics and astrophysics has been well known in the literature. We again refrain ourselves from these discussions here.

7.3 BERRY PHASE AND DYNAMICAL INVARIANTS

In 1984, Berry [203] made an important observation that any quantal system in an eigenstate transported adiabatically around a closed cycle attains, apart from the usual dynamical phase, a topological phase now known as Berry's phase. Berry phase and its applications have been investigated in different theoretical and experimental contexts like quantum Hall effect [204], fractional statistics of vortices in 2D [205], Born-Oppenheimer approximation [206] etc. Classical analogue in the form of Hannay angle [207] and semiclassical generalizations of Berry phase have also been looked into [207]. Several other features like gauge structure inherent in some simple quantum systems have also been studied [208] in terms of Berry's phase. Also, there have been discussions of Berry phase in the context of nonlinear systems [214]. For further details on Berry phase we refer to the book [209] and several other interesting works [208, 210].

The discussion on the Berry phase originally was in the context of adiabatic and time-periodic motions but now the same has been extended to the case of non-adiabatic and non-periodic motions by Aharonov and Anandan [211], Anandan [212] and by Berry [213] himself. It may be mentioned that while Berry phase has been investigated mostly for the 1D systems, not many attempts have been made so far to look for it in the 2D systems. As far as the connection of the Berry phase with the dynamical invariants is concerned it can be visualized through two seemingly different but intrinsically intimately connected [153] approaches, namely the time-evolution operator and dynamical Lie algebraic approaches discussed already in Sect. 5.2.2. Here we again refresh these approaches in the light of Berry phase and then make a few remarks concerning the role played by the dynamical invariants.

7.3.1 Berry Phase and Time-evolution Operator Approach

Several authors (see e.g. Refs. [210] and [215]) have followed the time-evolution operator (TEO) approach to derive the Berry phase in a quantum system described by the Hamiltonian $H(t)$ and the state function $|\,\Phi(t)>$; the latter has evolved from the initial state $|\,\Phi(0)>$ as $|\,\Phi(t)> = U(t)\,|\,\Phi(0)>$, where the TEO $U(t)$ with $U(0) = 1$ satisfies the TDSE (5.12) which can be expressed as

$$U^{\dagger}(t)\, H\, U(t) - i\, U^{\dagger}(t)\ \dot{}\ (t) = 0. \tag{7.58}$$

Further, we choose an orthonormal set of eigenstates $\{|\, n(0)>\}$ of $H(t)$ at $t = 0$ such that

$$H(0)\,|\, n(0)> = E_n\,|\, n(0)>, \tag{7.59}$$

where E_n is the instantaneous eigenvalue corresponding to $|n(0)>$, and then split the operator $U(t)$ in the form $U(t) = \mathcal{U}(t)\ \mathcal{R}(t)$. Use of this form in (7.58) gives rise to

$$\mathcal{U}^{\dagger}(t)\,[H - i\frac{\partial}{\partial t}]\,\mathcal{U}(t) \equiv \mathcal{H}(t), \tag{7.60}$$

where $\mathcal{H}(t)$ is the hermitian operator given by

$$\mathcal{H}(t) = i\, \dot{\mathcal{R}}(t)\, R^{+}(t). \qquad (7.61)$$

A change of representation relating $|\Phi(t)>$ to $|\Psi(t)>$ with the help of a unitary operator $\mathcal{U}(t)$ such that $|\Phi(t)> = \mathcal{U}(t)|\Psi(t)>$, further casts the evolution eq. (5.12) in the form

$$\mathcal{H}(t)|\Psi(t)> = i\frac{\partial}{\partial t}|\Psi(t)>, \qquad (7.62)$$

which is similar to the one satisfied by $|\Phi(t)>$ in conjunction with $H(t)$. However, eq. (7.62) is readily solvable provided a proper choice of $\mathcal{R}(t)$ is made. In fact, one can demand $\mathcal{H}(t)$ to be diagonal in the basis $\{|n(0)>\}$ and the same is possible if $\mathcal{R}(t)$ is diagonal in $\{|n(0)>\}$. As a result, the unitarity condition for $\mathcal{R}(t)$ will allow to express the matrix elements of $\mathcal{R}(t)$ in the basis $\{|(0)>\}$ as [210]

$$[\mathcal{R}(t)]_{mn} = \delta_{mn} \exp(-i\,\theta_n(t)), \qquad (7.63)$$

for some $\theta_n(t)$ satisfying $\theta_n(0) = 0$. With these specifications the operator $\mathcal{H}(t)$ is expressible as $[\mathcal{H}(t)]_{mn} = \delta_{mn}\theta_n(t)$. This diagonal representation of $\mathcal{H}(t)$ provides the solution of (7.62) and subsequently an expression for $|\Phi(t)>$ as

$$|\Phi(t)> = \mathcal{U}(t).\exp\left(-i\int_0^t \mathcal{H}(u)\,du\right)|\Phi(0)>, \qquad (7.64)$$

which describes the evolution of the quantal system.

If one starts with an eigenstate of H at time $t = 0$ such that $|\Phi(0)> = |n(0)>$, then one can obtain [210] $|\Phi(t)>$ in terms of the dynamic and nondynamic phases as

$$|\Phi(t)> = \exp(i\gamma_D(t)).\exp(i\gamma_B(t))\;|n(t)>. \qquad (7.65)$$

Here $\gamma_D(t)$ is the usual dynamic phase given by

$$\gamma_D(t) = -\int_0^t <\Phi(u)|H|\Phi(u)>du, \qquad (7.66a)$$

and $\gamma_B(t)$ is the nondynamic phase (similar to the one obtained by Berry [203]) expressible here as

$$\gamma_B(t) = i\int_0^t <n(u)|\dot{n}(u)>du, \qquad (7.66b)$$

where $|n(t)> = \mathcal{U}(t)|n(0)>$. Note that in obtaining (7.65) no adiabatic assumption is used unlike the one necessary in the derivation of this result by Berry [203].

tation of Berry phase in their work again however turn out to be difficult.

7.3.2 Berry Phase and Dynamical Lie Algebraic Approach

Moteoliva et al [216] (henceforth termed as MKN) have actually demonstrated the role of dynamical invariant operator in computing the Berry phase for a restricted class of TD systems. For this purpose a quantum analogue of the dynamical Lie algebraic approach, discussed already for the classical case in Chap. 3, is used to determine the invariant operator. The MKN approach while assumes the existence of an invariant operator $I(t)$ for the given TD system, offers a simplified version of the difficult TEO approach discussed in the previous subsection. In fact for such TD systems a relationship exists [216] between the approaches of Moore and Stedman [215] and that of Lewis and Riesenfeld [64] which helps in constructing both cyclic initial states and their associated Berry phases. In what follows we outline such a relationship.

Consider a quantum system described by a T-periodic Hamiltonian $H(t)$ and the state function $|\Phi(t)>$ evolves from the initial state $|\Phi(0)>$ as (cf. Sect. 7.3.1) $|\Phi(t)> = U(t)|\Phi(0)>$, where $U(t)$ with $U(0) = 1$ again satisfies the TDSE (5.12). Moore and Stedman [215] while carrying out Berry's formulation of the geometric phase for non-adiabatic evolution of such a system, decompose $U(t)$ in the form

$$U(t) = Z(t)\, e^{iMt}, \tag{7.67}$$

where $Z(t)$ is T-periodic unitary operator and M is a constant hermitian operator. Clearly, $U(0) = 1$ implies $Z(0) = 1$. Further note that such a decomposition of $U(t)$ is not unique and the cyclic initial state $|\Phi_m(0)>$, given by

$$|\Phi_m(T)> = U(T)|\Phi_m(0)> = \exp(i\chi_m)\,|\Phi_m(0)>, \tag{7.68}$$

is the eigenstate of M and returns to itself after a time T, with a phase χ_m. According to Moore and Stedman the non-adiabatic geometric phases associated with $|\Phi_m(0)>$ are given by

$$\gamma_m = i\int_0^T <\Phi_m(0)|\, Z^+(t)\dot{Z}(t)\,|\Phi_m(0)>dt. \tag{7.69}$$

MKN [216] assume that the system possesses a dynamical invariant operator $I(t)$ in accordance with eq. (5.17) and $I^\dagger(t) = I(t)$. Besides that $I(t)$ is one of a complete set of commuting observables, other conditions that it is assumed to fulfil are (i) the eigenstates $|n,t>$ of $I(t)$ are a complete set, (ii) $I(t)$ does not involve time-derivative operator, (iii) $I(t)$ is T-periodic, and (iv) its eigenvalues λ_n are non-degenerate. For such a system, following the work of Lewis and Riesenfeld [64] the solution of the SE can be constructed as (cf. eq. (5.2))

$$|\Phi(t)>=\sum_n C_n|\,\psi_n(t)>, \tag{7.70}$$

where $|\,\psi_n(t)> = \exp\,(i\alpha_n(t))\,|\,n, t>$ and the expansion coefficients C_n's do not depend on t. Further, the phases $\alpha_n(t)$ are determined from eq. (5.4). The assumption that $I(t)$ is T-periodic implies $|\,n, T> = |\,n, 0>$ and consequently a particular solution $|\,\psi_n(T)>$ (cf. eq. (7.70)) after one period becomes.

$$U(T)\,|\,\psi_n(0)> \equiv |\,\psi_n(T)> = e^{i\alpha_n(T)}|\,n, 0> = e^{i\alpha_n(T)}|\,\psi_n(0)>. \tag{7.71}$$

Thus, $|\,\psi_n(0)>$ are the eigenstates of $U(T)$ with eigenphases $\alpha_n(T)$.

Following the method of Moore and Stedman, one can choose $|\,\psi_n(0)>$ as the cyclic initial states and can avoid subsequently the use of evolution operator by writing $Z(t) = U(t)\,e^{-iMt}$, $|\,\psi_n(t)> = Z(t)\,|\,\psi_n(0)>$ in (7.69). This implies [216]

$$|\,\psi_n(t)> = U(t)\;e^{-iMt}\,|\,\psi_n(0)> = e^{-i\alpha_n(T)t/T}\,|\,\psi_n(t)>, \tag{7.72}$$

and

$$i<\Psi_n(T)\big|\dot{\Psi}_n(T)> = (\alpha_n(T)/T - \dot{\alpha}_n(T) + i<n,t\big|\,n^\circ,t>, \tag{7.73}$$

where $|\,n^\circ, t>$ denotes the time-derivative of the state $|\,n, t>$. Use of (7.72) and (7.73) in (7.69) yields the geometric phase in the form

$$\gamma_n(T) = i\int_0^T <n,t\big|n^\circ,t>dt, \tag{7.74}$$

for the initial states $|\,\psi_n(0)>$.

For the construction of the invariant operator $I(t)$ one follows the quantum analogue of the dynamical algebraic approach (cf. Sect. 3.2.3). For this purpose, one expands $H(t)$ in the form

$$H(t) = \sum_{i=1}^{N} h_i(t)\,\Gamma_i, \tag{7.75}$$

where the set $\{\Gamma_i\}$ generates a dynamical algebra which is now closed under the commutator (instead of its closure under the Poisson bracket as in the classical case)

$$[\Gamma_i, \Gamma_j\,] = \sum_{k=1}^{N} C_{i\,j}^{k}\,\Gamma_k, \tag{7.76}$$

with C_{ij}^k as the structure constants. Here Γ_i's do not depend explicitly on time and $h_i(t)$ are T-periodic. Since $I(t)$ is also a member of this algebra, so we have

$$I(t) = \sum_{i=1}^{N} a_i(t)\, \Gamma_i \,. \tag{7.77}$$

Further, the 'invariant' condition, $(d\,I\,(t)\,/\,dt) = 0$ (cf. eq. (5.16)), leads to a set of first-order ODE's for a_i's as

$$\dot{a}_i(t) = \sum_{j=1}^{N} A_{ij}(t)\,.\,a_j(t)\,, \qquad (i = 1, \ldots .\, N), \tag{7.78}$$

where $a_j(t)$'s are real since $I\,(t)$ is hermitian, and $A\,(t)$ is a T-periodic matrix with elements

$$A_{\ell k}(t) = -\frac{1}{i\hbar} \sum_j C_{\ell j}^{k}\, h_j(t).$$

Once the a_i's from (7.78) are known, then the operator $I\,(t)$ can be constructed from (7.77).

Using the above method, MKN [216] have computed the Berry phases for the cases of generalized harmonic oscillator described by the Hamiltonian

$$H(t) = \frac{1}{2}\,[\, x(t)\, q^2 + y(t)\,(pq + qp) + z(t)\, p^2 \,],$$

where q and p are the position and momentum operators, and of two-level system described by the Hamiltonian

$$H(t) = \frac{1}{2}\,[z\,|1\rangle\langle 1| - z\,|2\rangle\langle 2| + (x - iy)\,|1\rangle\langle 2| + (x + iy)\,|2\rangle\langle 1|\,].$$

Here x, y and z are the TD real parameters.

7.4 CONCLUDING DISCUSSION

In view of what has been presented above in this chapter, obviously it is not possible to assign physical meaning(s) to all invariants which exist and can be obtained for a dynamical system. From this point of view the well known case of Ermakov invariant $I\,(t)$ obtained for a TDHO system in 1D is discussed in detail. An interpretation for $I\,(t)$ due to Eliezer and Gray [170], as an angular momentum in a projected 2D plane, appears to be a plausible one. The same invariant when written in the quantum context [171-173] in the form of Ermakov-type functional invariant with respect to the position variable seems to have different interpretation.

From the list of applications in Sect. 7.2, it is evident that the role of invariants can not be ignored in some of the physical problems. In other cases although the invariants as such do not appear directly, nevertheless their scope in the problem is specified in the sense that an appropriate remodeling of the problem might accommodate them and subsequently make the understanding of the phenomenon better. Not only this if the invariants exist and are obtained for a system, then they can offer a deeper insight into the problem. As a matter of fact the difficulties in dealing with invariants in a physical problem arise at several stages: First, at the stage of existence of the invariant. There is no general criterion at present which can predict the existence of invariants of a system in advance. By and large they are obtained either by a hit-and-trial method or by accident. Secondly, at the stage of their construction. Very often different orders in momenta (for polynomial invariants) have to be tried to obtain finally only few. Sometimes the available methods appear inadequate for this purpose. Lastly, comes the stage of interpretation (which is possible only in rare cases) and then is the stage of application to a physical problem.

8

Constrained Dynamical Systems and Invariants

8.1 INTRODUCTION

Whether it is classical (cf. Chaps. 2 and 3) or quantum context (cf. Chaps. 4 and 5) so far we have talked of the systems which are free from additional restrictions (called constraints) on the canonical variables of the system. For this reason only we have freely used the Poisson brackets in the classical context and the commutators in the quantum context as and when their discussion is required. In the presence of constraints, however, some modifications at the foundation level itself are necessary. In this chapter, we aim at a brief discussion of these modifications in the context of both classical and quantal ideas.

As a matter of fact the quantization of classical systems in the presence of constraints has been a subject of study for a long time. Starting with the work of Dirac [219, 220] this subject in modern times has very widely been discussed in the textbooks (see, for example, Sudarshan and Mukunda [218]) and also in several other works [221, 222]. Dirac recognized that a consistent way to deal with the situation is the naive modification of the Poisson brackets such that the new brackets (known as the Dirac brackets) between a physical variable and a constraint vanish.

Note that in the construction of invariants we basically consider the time evolution (linked with the Poisson bracket in the classical case (cf. eqs. (1.13) and (1.14)) and with the commutator in the quantum case (cf. eq. (5.16)) besides the fact that the physical quantities like coordinates or momenta or their functions now become operators in the Hilbert space in the latter case) of a phase space function. As far as the role of constraints in the context of the present book is concerned, following Dirac [219, 220] we just have to highlight their effect on the time evolution of a phase space function $F(q,p)$ or on the concerned Poisson bracket related by

$$\frac{dF}{dt} = [F,H]_{\text{PB}}. \tag{8.1}$$

In other words, we plan to find the effect of superfluous degrees of freedom which a system attains in the presence of external conditions (and known as constraints) on relation (8.1). From this point of view the study of singular dynamical systems (in the sense of nonstandard Lagrangians, see next section) has become a matter of immediate concern mainly because such systems turn out to be of interest in the context of field theories. In the next section we shall just suggest (rather than working out precisely for particular examples) some possible guidelines towards the improvements of the existing methods for the construction of exact invariants in the presence of constraints. These discussions will be continued over to the quantum domain in Sect. 8.3, with some concluding remarks in the context of field theory in Sect. 8.4.

8.2 ROLE OF CONSTRAINTS IN THE CLASSICAL CONTEXT

The role of constraints in the classical context can be investigated in both Lagrangian and Hamiltonian formulations of the problem. Since the Hamiltonian formalism is more akin to performing the quantization in phase space variables in the quantum context (cf. next section), we therefore emphasize here more on this formalism than on the Langrangian one.

Following Dirac [220] we consider a dynamical system described in terms of coordinates q_n $(n = 1, 2, .. N)$ and velocities $\dot{q}_n$, with a Lagrangian $L(q_n, \dot{q}_n, t)$ related to the Hamiltonian $H(q,p)$ through the Legendre transformation

$$H(q,p) = \sum_n p_n \dot{q}_n - L(q_n, \dot{q}_n, t), \tag{8.2}$$

where the momenta p_n are defined by

$$p_n = (\partial L / \partial \dot{q}_n). \tag{8.3}$$

It may be that the p's are not independent functions of $\dot{q}$'s. If p's involve only $(N - M)$ independent functions of $\dot{q}$'s, there will be M complete set of independent M relations (say) of the type

$$\Gamma_\alpha(q,p) = 0, (\alpha = 1,2, ... M), \tag{8.4}$$

with $0 < M < N$. Before proceeding further we introduce a few terms here.

Standard and nonstandard Lagrangians: Note that the Legendre transformation (8.2) implies the passage from $\{q_n(t), \dot{q}_n(t)\} \rightarrow \{q_n(t), p_n(t)\}$, which allows us to write

$$\frac{\partial p_n}{\partial \dot{q}_\ell} = \frac{\partial^2 L(q_n, \dot{q}_n)}{\partial \dot{q}_\ell\, \partial \dot{q}_n}. \tag{8.5}$$

The right hand side of (8.5) represents the elements of an $N \times N$ matrix W as $W_{\ell n} = (\partial^2 L\,(q_n, \dot{q}_n) / \partial \dot{q}_\ell\, \partial \dot{q}_n)$. Now we look at this matrix W as the Jacobian of the transformation. If det $(W_{\ell n}) \neq 0$, i.e. the matrix W is nonsingular, then the corresponding Lagrangian is called standard, otherwise it is called nonstandard. In the same way one can define [218] standard and nonstandard Hamiltonians by starting with $\dot{q}_n = (\partial H / \partial p_n)$, but only after keeping in mind the several points of distinction in this case. In terms of W, the Euler-Lagrange equations of motion (1.16) now take the form

$$W_{\ell n}\, \ddot{q}_n = K_\ell(q, \dot{q}), \tag{8.6}$$

where

$$K_\ell(q, \dot{q}) = (\partial L / \partial q_\ell) - (\partial^2 L / \partial \dot{q}_\ell\, \partial q_n)\dot{q}_n.$$

By setting $\dot{q}_n = \upsilon_n$, one can also see the appearance [222] of the matrix W in a set of first order equations. Further the variation of the momentum p_ℓ now defined by $p_\ell = (\partial L(q, \upsilon) / \partial \upsilon_\ell)$, in the form

$$\delta p_\ell = W_{\ell n}\, \delta \upsilon_n + (\partial^2 L / \partial q_n \partial \dot{q}_\ell)\, \delta q_n, \tag{8.7}$$

implies that if the matrix W is singular then $\delta \upsilon_n$ cannot be expressed in terms of δp_ℓ and δq_n.

For the case of singular W some momenta depend on others and this corresponds to a singular Lagrangian. Further let R be the rank of the Jacobian (8.5) with $R < N$, then the $N \times N$ matrix W can be recast into a form with diagonal submatrices of dimensionality R and $(N - R)$ such that $W_{\alpha\beta} = (\partial^2 L / \partial \dot{q}_\alpha\, \partial \dot{q}_\beta)$ for $\alpha, \beta = 1, \ldots, R$ has det $(W_{\alpha\beta}) \neq 0$ and for $\alpha, \beta = R+1, \ldots, N$ has det $(W_{\alpha\beta}) = 0$. Accordingly the defining eq. (8.3) can also be rewritten [218] for these two situations. In fact, Γ_α must be independent functions of the p's by themselves, otherwise one can eliminate p's from (8.4) and obtain constraints involving only q's. This, however, is not acceptable at this stage; instead without any loss of generality, suppose that eq. (8.4) allows us to express $p_{R+1}, \ldots, p_M$ in terms of p_1, ... ,p_R and the q's as

$$p_\rho = \psi_\rho(q_n, q_\alpha), \ (\rho = R+1, \ldots, M;\ \alpha = 1, \ldots, R), \tag{8.8}$$

where ψ's are the functions of their arguments.

Weak and strong equalities: Note that any departure from the defining eq.

(8.3) (say by an order of $\in$) allows us to introduce the concept of weak and strong equalities. In order to understand this concept consider a 3N dimensional space with q's, $\dot{q}$'s and p's as coordinates. In this space eq. (8.3) will define $2N$ dimensional region (say $\mathcal{R}$). In view of the fact that in this region the constraints (8.4), which basically are the consequences of (8.3), will also be satisfied and thereby further limit the region $\mathcal{R}$. In fact, all points of the $3N$ dimensional space which are within a distance of order of $\in$ from $\mathcal{R}$ will form a $3N$ dimensional region (say $\mathcal{R}_\in$) like a shell of thickness of order $\in$. A weak equality (denoted by "$\approx$") holds in the region $\mathcal{R}$ whereas a strong equality (denoted by "$\equiv$") holds in the region $\mathcal{R}_\in$. For example, two functions A, B of dynamical variables $q_n(t)$, $p_n(t)$, $n = 1, .. , N$, are said to be weakly equal i.e. $A\ (q, p) \approx B\ (q, p)$ if they are equal on the hypersurface Γ_c defined by the (primary) constraints (8.4) i.e. *if* $(A\ (q, p) - B\ (q, p))\ |_{\Gamma_c} = 0$. On the other hand, they are said to be strongly equal if not only A equals B on Γ_c but also the gradients of A (i.e. $(\partial A / \partial q_n,\ \partial A / \partial p_n)$) agree with that of B (i.e. $(\partial B / \partial q_n,\ \partial B / \partial p_n)$) when the arguments are restricted to Γ_c.

Now let us make arbitrarily small variations δq_n, $\delta \upsilon_n$ which give rise to the variation δp_n as in (8.7) and preserve the definition of p_n (cf. eq. (8.3)). These variations must also preserve eqs. (8.4) and subsequently imply

$$(\partial \Gamma_\alpha / \partial q_n)\, \delta q_n + (\partial \Gamma_\alpha / \partial p_n)\, \delta p_n = 0. \tag{8.9}$$

Notice that if constraints (8.4) are written in such a way that first order independent variations in the p's and q's make first order variations in Γ's then eq. (8.9) is the only restriction on δp_n.

Next consider the variation of the quantity $(p_n \dot{q}_n - L)$ as

$$\delta(p_n \dot{q}_n - L) = p_n\ \delta\dot{q}_n + \dot{q}_n\ \delta p_n - (\partial L / \partial q_n)\ \delta q_n - (\partial L / \partial \dot{q}_n)\ \delta \dot{q}_n,$$

which, after using the definitions of H and p_n, respectively from (8.2) and (8.3), reduces to the form

$$\delta H = \dot{q}_n\, \delta p_n - (\partial L / \partial q_n)\, \delta q_n. \tag{8.10}$$

From (8.10) it can be noticed that (i) δH does not depend on $\delta \dot{q}_n$'s and hence leaves the quantity $(p_n \dot{q}_n - L)$ unchanged w.r.t. the variations in $\dot{q}$'s, and (ii) H turns out to be the function of q's and p's only and as a result of (i) it is not determined uniquely. One may change H by

$$H \to H^{(1)} = H + \sum_\alpha \lambda_\alpha \Gamma_\alpha, \tag{8.11}$$

with λ_α's (some kind of Lagrange multipliers) being functions of q's and p's.

A comparison of variations in H from (8.10) and that in $H^{(1)}$ will yield

$$\dot{q}_n = (\partial H / \partial p_n) + u_\alpha (\partial \Gamma_\alpha / \partial p_n), \tag{8.12}$$

$$-(\partial L / \partial q_n) = (\partial H / \partial q_n) + u_\alpha (\partial \Gamma_\alpha / \partial q_n), \tag{8.13}$$

where u_α are some suitable coefficient functions [220]. The use of equation of motion (1.16) will reduce (8.13) to the form

$$\dot{p}_n = -(\partial H / \partial q_n) - u_\alpha (\partial \Gamma_\alpha / \partial q_n). \tag{8.13'}$$

After using these generalized Hamilton's equations of motion, the time evolution of a phase space function $F(q,p)$ in the presence of constraints (8.4) can immediately be written as

$$\frac{dF}{dt} = [F,H]_{PB} + u_\alpha [F,\Gamma_\alpha]_{PB}, \tag{8.14}$$

where the second term on the right hand side (RHS) is the modification to eq. (8.1) arising solely from the constraints. The RHS of (8.14), sometimes designated as "Dirac bracket" and denoted by $[.\,,.]_D$ is the generalization of the Poisson bracket and has many properties in common with it. Further, following remarks about (8.14) are in order:

(i) Note that H through its dependence on q's, $\dot{q}$'s and p's belongs to the region $\mathcal{R}_\epsilon$ whereas $H^{(1)}$ through its dependence on q's and p's only belongs to $\mathcal{R}$. This implies a weak equality in (8.14).

(ii) As far as the construction of an exact dynamical invariant I is concerned it can still be obtained from $(dI/dt) = 0$ as before except for the fact that (dI/dt) is now given by (8.14) instead of (1.13). This "invariant" condition through

$$[I,H]_{PB} + u_\alpha [I,\Gamma_\alpha]_{PB} = 0, \tag{8.15}$$

will imply (perhaps) further restrictions on the constraints Γ_α in order that the invariant exists for the system. Moreover, the exactness of the invariants so constructed remains somewhat questionable in the sense of prevailing weak equalities in the presence of constraints.

(iii) If we differentiate (8.4) w.r.t. t and use (8.14) for the time evolution of Γ_α, we get

$$[\Gamma_\beta, H]_{PB} + u_\alpha [\Gamma_\beta, \Gamma_\alpha]_{PB} \approx 0. \tag{8.16}$$

This equation (unless it reduces to an identity) will provide further constraints and help in reducing the number of independent Hamiltonian variables q's, p's and u's. While these relations between q's and p's only may turn out to be independent of the Γ_α equations (cf. eq. (8.4)), they correspond to the weak equalities of the type

$$\chi_\sigma(q,p) \approx 0, \quad (\sigma = 1, 2, \ldots). \tag{8.17}$$

Note that the differentiation of (8.17) w.r.t. t will again be expressible from (8.14) as

$$[\chi_\sigma, H]_{PB} + u_\alpha[\chi_\alpha, \Gamma_\alpha]_{PB} \approx 0. \tag{8.18}$$

If (8.18) is not again an identity, then it will imply another set of relations among q's and p's and subsequently one can go on repeating this process and the use of (8.14), as far as it goes i.e. until all the constraints are determined to be evolution free.

(*iv*) Note that Γ_α's appear in the general equation of motion (8.14). In literature Γ_α's are termed as "primary" constraints whereas the constraints of χ_σ-type are called "secondary". For many purposes Γ_α's and χ_σ's are treated on the same footing. Not only this, on the basis of constraints there exists the classification of dynamical variables or phase space functions. For example, a function (say R) of q's and p's is said to be "first class" if its Poisson bracket with H and with all constraints (including primary and secondary) vanish even weakly (in view of (8.17) or like that); otherwise R is called "second-class". If R is first-class, then $[R,\Gamma_\delta]_{PB}$ (where Γ_δ's now include Γ_α's and χ_σ's) is strongly equal to some linear function of Γ_δ's. For further details we refer to the literature [219, 220].

(*v*) By looking at the similarity of eqs. (8.15), (8.16) and (8.18) one can get the feeling that the invariants are also constraints. But this is not so, rather invariants carry much deeper meaning than merely appearing as the phase space functions. We avoid details here and again refer to literature [218-220].

(*vi*) For the constrained systems involving explicit time dependence the above discussions hold more or less as such but for an additional term $(\partial F/\partial t)$ on the RHS of (8.14).

8.3 ROLE OF CONSTRAINTS IN THE QUANTUM CONTEXT

In Chaps. 4 and 5 our emphasis was mainly on the use of Schrodinger wave mechanics to study the problems related to NC potentials. To the best of our knowledge the role of constraints in this regard has not yet been discussed in literature. Instead, from the survey given in the previous Section it appears that the role of constraints in the quantum context can be better understood through Heisenberg (matrix) mechanics where the commutators of the dynamical variables are used. In fact, as mentioned before, the use of commutators as the "simple" (?) extension of the corresponding Poisson brackets (or of the Dirac brackets of the previous section for that matter) to the quantum domain, has also been the subject of study for a long time as far as the quantization of constrained dynamical systems is concerned. In what follows, we briefly discuss the canonical quantization of constrained systems with a view to highlight the underlying difficulties in this respect.

By definition the canonical qunatization consists of obtaining the Poisson bracket relation between any two physical variables and subsequently carry them over to the quantum commutation relation (or anticommutation relation which are more pertinent in the case of field theories and we avoid the discussion of the same here) with an $(i\hbar)$ prescription. For example, for two physical variables A and B this procedure implies

$$[\hat{A}, \hat{B}] = i\hbar\, [A,B]_{\mathrm{PB}}. \tag{8.19}$$

In quantum mechanics, since $[\hat{q},\hat{p}] - i\hbar\, [q,p]_{\mathrm{PB}} = i\hbar$, this method while works quite well, it poses several difficulties not only in the generalized version of quantum mechanics (particularly, when the constraints are present) but also at the level of quantum field theory. As a matter of fact, according to this prescription, in the stage of passing to the quantum theory, a constraint of the type (8.4) must map to a null operator $\hat{\phi}$, namely $\Gamma(q,p) \to \hat{\phi}$, which subsequently implies

$$[\hat{A}, \hat{\Gamma}] = [\hat{A}, \hat{\phi}] = i\hbar\, [A, \Gamma]_{\mathrm{PB}}. \tag{8.20}$$

Note that while LHS of this equation clearly vanishes as it is the commutator of an operator with the null operator, the RHS does not, as the Poisson bracket of A and Γ is not in general zero. So this is the kind of inconsistency which appears in dealing with constraints in the quantum context.

To some extent this inconsistency is overcome by using the Dirac bracket in place of the Poisson bracket in (8.20), viz.,

$$[\hat{A}, \hat{B}] = i\hbar\, [A, B]_D. \tag{8.21}$$

where [218]

$$[A, B]_D = [A, B]_{\mathrm{PB}} - \sum_{\alpha,\beta} [A, \Gamma_\alpha]\, C^{-1}_{\alpha\beta}\, [\Gamma_\beta, B]_{\mathrm{PB}}, \tag{8.22}$$

with $C_{\alpha\beta} \approx [\Gamma_\alpha, \Gamma_\beta]_{\mathrm{PB}}$ i.e. the elements of the matrix formed by the Poisson brackets between second class constraints. Clearly, the matrix $(C_{\alpha\beta})$ is antisymmetric and also nonsingular so that its inverse exists.

As far as the existence and subsequently the determination of an invariant operator $\hat{I}$ corresponding to a given Hamiltonian operator $\hat{H}$ in two and higher dimensions is concerned, one can proceed through the Heisenberg equation of motion for a TID system. On the other hand, one can use time evolution operator approach (cf. Sect. 5.2.2) for the TD case. The former will require that $[\hat{I}, \hat{H}] = 0$, for $\hat{I}$ to be an invariant operator, which in turn using the prescription (8.21) implies $[I,H]_D = 0$, or

$$[I,\ H]_{PB} - \sum_{\alpha,\beta} [I,\ \Gamma_\alpha]_{PB}\ C^{-1}_{\alpha\beta}\,[\Gamma_\beta,\ H]_{PB} = 0. \qquad (8.23)$$

Eq. (8.23) represents an alternative form of eq. (8.15) corresponding to the so called quantum case.

8.4 CONCLUDING DISCUSSION

It may be mentioned that the brief sketches given in the previous sections are valid not only for the finite dimensional systems but also for the infinite dimensional ones (as is the case in field theory). As a result, the handling of the NC dynamical systems in the presence of constraints is not a problem as such in the context of classical and quantum mechanics. In the quantum domain, however, the difficulty arises from another front with regards to the construction of invariants and that is in terms of the connection between the commutator and the Poisson bracket (or the Dirac bracket for that matter) in general. As mentioned before (and also see Appendix 2 for the details) in the classical limit ($\hbar \to 0$) a commutator does not reduce to the Poisson bracket as such, particularly when the invariants of order higher than two (in momenta) are involved. In that case, in fact, the quantum corrections arise and their account becomes necessary in order that a classical invariant is also a quantum invariant. Since the invariants basically are the phase space functions with any type of its dependence on q's and p's in general, such corrections are bound to appear. Thus, the construction of an exact invariant in the presence of constraints may be simpler at the classical level than that at the quantum level. Further, at the quantum level the Heisenberg formulation could be more appropriate for this purpose.

An area of frequent applications of the theory of constraints is the subject of field theory. The typical examples often discussed in the literature [221-223] are that of the electromagnetic field described by the Lagrangian $L = -(1/4)\,F_{\mu\upsilon}\,F^{\mu\upsilon}$, where $F_{\mu\upsilon} = \partial_\mu\, A_\upsilon - \partial_\upsilon A_\mu$, with $\mu, \upsilon = 0, 1, 2, 3$ and of the nonlinear sigma model in 1+1 dimensions described by $L = (1/2)\ \partial_u\ \varphi^i\ \partial^\mu\ \varphi^i s$, $(i = 1, \ldots, n)$. In the former case, the conjugate momentum field $\Pi_\mu(x) = (\partial L / \partial \dot{A}_\mu)$, while expressed in the component form as $\vec{\pi}(x)\,(= (\dot{\vec{A}} + \vec{\nabla}\, A^\circ))$, $\pi^\circ(x)$, satisfy the (first class) constraints $\pi^\circ(x) \approx 0$, $\vec{\nabla}.\vec{\pi}(x) \approx 0$. Accordingly the gauge is fixed and one convenient gauge for the electromagnetic field is the Coulomb gauge $\vec{\nabla}.\vec{A}(x) \approx 0$. On the other hand, the constraint used in the second case is $\varphi^i\varphi^i = 1$ which can be incorporated by rewriting the Lagrangian as

$$L = \frac{1}{2}\partial_\mu\varphi^i\,\partial^\mu\varphi^i + \frac{\lambda}{2}(\varphi^i\varphi^i - 1), \tag{8.24}$$

which possesses a primary constraint $\pi_\lambda = (\partial L / \partial\dot{\lambda}) \approx 0$.

In order to understand further the role of constraints, we present here the classical analogue of the nonlinear sigma model i.e. of the second example mentioned above. Consider the motion of a particle of mass m constrained to move freely on a sphere of radius unity and described by the Lagrangian (analogous to (8.24))

$$L = \frac{1}{2}m\sum_i \dot{x}_i^2 + \lambda\ (\sum_i x_i^2 - 1). \tag{8.25}$$

By treating λ as a dynamical variable the equations of motion for x_i and λ turn out to be

$$m\,\ddot{x}_i - 2\lambda\, x_i = 0,\ \sum_i x_i^2 - 1 = 0. \tag{8.26a,b}$$

Note that by differentiating (8.26b) twice w.r.t. t and comparing the resultant equation with eq. (8.26a), one can obtain λ (after using eq. (8.26b) again) as $\lambda = -(1/2)m\sum_i \dot{x}_i^2$. Subsequently, the equation of motion (8.26a) can be recast as

$$m\,\ddot{\vec{x}} - (m\,\dot{\vec{x}}^2)\vec{x} = 0, \tag{8.27}$$

which corresponds to a system involving the velocity dependent potential. Thus, the presence of constraints sometimes changes the nature of the original system or suggests us to consider an alternative constraint-free system which perhaps is more complex than the original one from the point of view of its physical understanding.

9

Summary and Future Prospects

Within the scope of the book, as spelt out in Chap. 1, a survey of classical and quantum mechanics of both TID and TD systems in 1D and 2D is carried out. Further it may be mentioned that as far as possible all throughout, whether it is classical or quantum context, exact invariants which are polynomials in momenta are dealt with. The discussion of approximate invariants (about which enormous literature has been known for a long time now, particularly for TD systems) is, however, wishfully avoided. Although the concluding remarks pertaining to these aspects of dynamical systems are made in each chapter separately still an essence of the contents of the monograph is worth presenting here at one place. This is done in the next Section.

In order to look at this limited survey in relation to other recent advancements in the study of dynamical systems, a general discussion is presented in Sect. 9.2. Here, again some passing remarks will be made on those related aspects of dynamical systems which are not covered in the book but appear to be the thrust areas in the present times. Finally, since surveys of such ever growing subjects can never in principle be up-to-date, therefore, it becomes natural to highlight once again (inspite of their glimpses given now and then in the text) in Sect. 9.3, not only the ongoing but also the future problems to be tackled with respect to the pace of time.

9.1 SUMMARY

After providing with enough preliminary background mainly from the classical mechanics in Chap. 1, the construction of exact invariants for TID systems is carried out in the subsequent chapter. As several of such systems have been the subject of study for centuries and many of them for decades now, only a selective survey is presented in this regard. With a view to obtaining new integrable systems (by way of constructing the second invariant besides the Hamiltonian) the complexification method is employed to study the 2D systems. This method, based on the use of the complex coordinates $Z = x_1 + ix_2$ and $\bar{Z} = x_1 - ix_2$ in place of the Cartesian ones x_1 and x_2, has not only reproduced many of the earlier results but also provided several new integrable systems.

In the classical context our survey is mainly directed to the study of so far less explored field of dynamical systems which involve explicit time-dependence. An exhaustive study presented in Chap. 3 not only includes the survey of integrable 1D systems (by way of constructing the only invariant in this case) but also an up-to-date survey of TD systems in 2D. In the latter case various methods for the construction of first invariant are presented including the complexification method once again and the dynamical algebraic approach. The complexification method in this case is found useful in providing second order invariants for a number of TD central potentials. Also, an extension of this method to the case of third order invariants is outlined to the extent that the solution of the derived 'potential' equations will directly provide the systems admitting such invariants. On the other hand, the uses of rationalization method (cf. Sect. 3.3.1) and of dynamical algebraic approach suggest a new class of Ermakov systems in 2D. Not only this, using these methods at least one exact invariant is constructed for a variety of TD systems in 2D. Attempts are also made to look for a method which can provide the second invariant for such systems. In general, the study of a TD system turns out to be more difficult than that of a TID system for the given dimension.

In the quantum context, exact closed-form solutions of the SE are investigated for NC potentials in 2D and for both TID and TD systems in Chaps. 4 and 5, respectively. In the TID case, while difficulties in dealing with NC AH potentials in general are explicitly discussed, normalizable closed-form solutions corresponding to nonzero eigenvalues are given for a number of NC potentials in 2D. In general, it is observed that this becomes possible if the parameters of the potential satisfy certain constraining relation(s). If the physical conditions fulfil these constraints on the couplings then the solutions obtained are exact. (These solutions are sometimes called quasi-exact as they correspond only to a few eigenvalues of the spectrum). A variety of power, exponential and parabolic NC potentials in 2D are studied using various methods. The role of constraints on potential parameters, while discussed for some typical potentials, turns out to be important, particularly for investigating the excited states of the system. It is suggested that the dynamical symmetries in the nonrelativistic quantum mechanics, if found for a given system, can also play an important role with regard to providing the exact solutions of the SE for that system.

An extension of the eigenfunction-ansatz method which is used to study

TID systems in 1D and 2D in Chap. 4, is now carried out to the TD case in Chap. 5, but with certain modifications. Unlike the TID case of Chap. 4, here the attempts are made to obtain the full spectrum of eigenvalues. In fact, it turns out that NC AH potentials pose certain problems as far as the study of exact solutions of the TD SE is concerned. On the other hand, a generalized TD HO potential in 2D is found to admit an exact solution of the SE. A brief survey of the formal approaches, like the time evolution operator, Lie algebraic and Feynman propagator approaches to study the quantum mechanics of TD systems, is also given (cf. Sect. 5.2.2 and 5.3). Such studies thus far are however restricted only to 1D systems.

In Chap. 6, mainly the extensions of various methods used for the 2D systems in Chaps. 2 to 5 are demonstrated to the case of 3D systems. In particular, the extension of rationalizaion method, method of Lie symmetries and dynamical algebraic approach in the classical context are discussed. It may be mentioned that while these extensions appear trivial in most cases, a study of 3D or higher dimensional systems using these methods, however, turns to be difficult for two reasons. Firstly, in the sense that the number of underlying equations to be dealt with now becomes large even for the case of second order invariants. Secondly, in the classical context, as pointed out in Sect. 6.3, the occurence of the phenomenon of 'Arnold diffusion' cannot be ruled out if the system under study turns out to be nonintegrable—a situation which is very likely. Also, for 3D systems the classical aspect has been studied much more in literature than the corresponding quantum aspect as far as their NC nature is concerned.

Practical applications and the role of dynamical invariants constructed in Chaps. 2 and 3 are highlighted in Chap. 7. Not only the possible interpretations advanced to some of the constructed invariants (particularly to TD invariants of Ermakov-type) were presented in this chapter but also the scope of these invariants in various branches of physics, chemistry and biology is demonstrated. In fact, under highly restricted conditions there seems to exist a connection between the TD dynamical invariants and the dynamical part of the Berry's phase. Finally, in the foregoing chapter the role of constraints in the process of construction of dynamical invariants is discussed.

9.2 A GENERAL DISCUSSION OF DYNAMICAL SYSTEMS IN MODERN CONTEXT

In this Section we list some of the thrust areas of modern times where in the study of quantum aspects of the system, TD invariants can play some role in varying proportions, namely (i) if the system is integrable, then the techniques discussed in this book can be used as such to study the time evolution of the system, (ii) if the system is nearly integrable in the sense that the unperturbed part is integrable and the TD perturbation is not, then these techniques can still be used but only marginally, and (iii) in the case when the system is not integrable, a somewhat different description (which is not described in this book) of the system is required. Within this framework we present below a brief discussion of the (i) time-periodic, and (ii) mesoscopic systems mainly from the quantal point of view. Recall that about time-periodic systems a few passing remarks were also made in Sect. 7.3.

Further note that in the literature [58]a TD problem is also converted into a TID one by considering physical time as a coordinate conjugate to the energy. In other words, the concept of an extended phase space in the context of Hamiltonian formalism has been introduced. As mentioned before (cf. Sect. 3.1) while this has been done mainly for the sake of mathematical abstraction, we have however avoided all throughout such a discussion and chosen instead to consider the dependence on time of the system rather explicitly. Here, we are going to take recourse to these considerations mainly for having an effective discussion of the derived results at a later stage.

9.2.1 Time-periodic Systems and Floquet Operator

For a TD Hamiltonian acting on a Hilbert space $\mathcal{H}$ the solutions of the TDSE (5.1) may be expressed through $\Psi(t) = U(t, s)\,\Psi(s)$ (cf. Sects. 5.2.2. and 7.3), where s is some initial time which is taken to be zero in earlier discussions and the TEO $U(t, s)$ is now given by [224] the integral equation (instead of the differential form (5.12))

$$U(t,s) = 1 - \frac{i}{\hbar}\int_s^t d\tau\, H(\tau)U(\tau,s). \tag{9.1}$$

Since we consider the physical time t as a coordinate conjugate to the energy E, the new Hamiltonian can be written as

$$K = H(t) + E, \tag{9.2}$$

and the corresponding Hamilton equations of motion with respect to the new fictitious time η become

$$\dot{t} = \frac{\partial t}{\partial \eta} = \frac{\partial K}{\partial E} = 1;\ \dot{E} \equiv \frac{\partial E}{\partial \eta} = -\frac{\partial K}{\partial t} = -\frac{dH}{dt}. \tag{9.3}$$

The latter equation in turn implies $(\partial K/\partial\eta) = (dH/d\eta)$. Note that the time η differs from t by the addition of a constant. In fact, η is the time parameterizing the Hamiltoninan flow, and the conservation of K requires that $(dK/d\eta) = 0$.

In the quantum context, using [225] $E = -i\hbar\,(\partial/\partial t)$ in eq. (9.2) one obtains $K = H(t) - i\hbar(\partial/\partial t)$ and the corresponding SE now becomes

$$K\Phi = i\hbar\frac{\partial\Phi}{\partial\eta}. \tag{9.4}$$

If the solutions of (5.1) belong to the Hilbert space $\mathcal{H}$, the solutions of (9.4) then lie in the extended Hilbert space $\mathcal{L}^2(\mathcal{R}, \mathcal{H})$ of $\mathcal{L}^2$-functions in $\mathcal{R}$ with values in $\mathcal{H}$. The operator $\exp(-i\eta K/\hbar)$ is related to the TEO of the TDSE (5.1) by

$$[\exp(-\frac{i}{\hbar}\eta K)(\Phi)](t) = U(t,t-\eta)\,\Phi(t-\eta). \tag{9.5}$$

Now, let $H(t)$ be time-periodic with period T i.e. $H(t+T) = H(t)$, then $U(t+T, s+T) = U(t, s)$ implies $U(t + NT, s) = U^N(t+T, s)$ and this reduces the problem to finding the TEO within one period. The one-period TEO $U(t+T, t) \equiv F_t$ is called *Floquet operator* in the literature [224-229].

As far as the spectral properties of the Floquet operator F_t are concerned, they do not depend on t because of the unitary equivalence $F_s = U(s,t)\, F_t\, U^{\dagger}(s, t)$, and K commutes with one-period translation operator. Also, in the extended phase space formalism one may confine [225] to the subspace $\bar{\mathcal{H}} = \mathcal{L}^2([0,T],\mathcal{H})$ of periodic functions with boundary conditions $\Phi(T) = \Phi(0)$. Such a restriction of the K-operator to this subspace leads to what is known as *quasi-energy operator* (K_T). Noting the fact that $U(t, 0)$ and $\exp(-itK_T/\hbar)$ are unitarily equivalent [226], the spectrum of the Floquet operator (or equivalently that of quasi-energy operator) is called the quasi-energy spectrum. From the spectral decomposition of $U(t, s)$ in the form [224]

$$(t,s) = \int e^{-i\lambda(t-s)}\, dP_\lambda, \quad (\lambda \in [0,2\pi]),$$

one can also obtain the wave function $\Psi(t)$ for any given initial condition. In other words the quasi-energy spectrum contains all the information needed for solving the quantum problem. As a matter of fact the nature of the quasi-energy spectrum and in particular the question of stability of the point spectrum has been investigated for several cases. A number of Hamiltonian systems which are quadratic in x and p and with time-periodic coefficients as well as with time periodic perturbations are discussed by Mendes [224].

Using the fact that in the extended phase space formalism a TD system in 1D can be considered as a TID system in 2D, Korsch and his coworkers in a series of papers [227-229] have applied the EBK quantization [141, 230] to study the semiclassical version of quasi-energies for a number of systems. They use the EBK quantum conditions in this case as

$$J_i = \frac{1}{2\pi}\oint_{\gamma_i} \vec{p}.\vec{dq}$$

$$= \frac{1}{2\pi}\oint_{\gamma_i} (p\,dq + E\,dt) = \hbar\,(n_i + \frac{1}{4}\mu_i), \quad (i = 1, 2), \qquad (9.6)$$

where n_1, n_2 are integers, γ_i denotes two independent closed paths on the torus which cannot be homotopically deformed into each other and μ_i are the Maslov indices of the paths. For details we refer to Ref. [227] and the references therein. In particular, the cases of forced quartic oscillator [228] (a special case of the classical Duffing Hamiltonian which shows both the regular and chaotic behaviour)

$$H(t) = (p^2/2m) + b\,x^4 - f\cos\phi\cos(\omega t), \qquad (9.7)$$

and that of a harmonically driven rotator [229]

$$H(t) = (p_\phi^2/2\mathcal{J}) - f\cos\phi\cos(\omega t), \qquad (9.8)$$

are studied. Here $\mathcal{J}$ is the moment of inertia and other symbols have their usual meanings. The fact that the numerical results obtained for quasi-energies of these systems turn out to be in good agreement with the exact quantum results for the regular states, implies the elegance of this semiclassical procedure. Further, for the systems which exhibit mixed (chaotic and regular) dynamics, Mirbach and Korsch [229] introduce the concept of quantum phase entropy. This concept might turn out to be useful for future studies of such systems.

9.2.2 Mesoscopic Systems

Some of the semiclassical concepts mentioned above have also become the subject of central discussion in recent years with reference to new experiments in atomic, molecular, microwave and mesoscopic physics. As a matter of fact the accuracy of the semiclassical approximations has strong bearing on the nature of classical trajectories. For the systems exhibiting chaotic behaviour at the classical level, these trajectories turn out to be extremely complex and thereby pose several questions with regard to either the validity or the very definition of these semiclassical approximations.

For the mesoscopic world, caught between classical and quantum behaviour, these semiclassical ideas are a necessity. Many experiments with excited atomic and molecular systems now directly reveal the realm of high quantum numbers or of classically chaotic motion. Further, the relation between the classical and quantum domains manifests through the connection that exist between the geometric ray paths and the wave phenomena associated, respectively, with these domains. While these connections could be of importance in many fields (like optics, acoustics, plasma physics, microwave physics etc.) which essentially involve identical problems, they always regard the geometrical limit as a useful approximation to the wave behaviour. Without going into details of the foundational aspects of quantum mechanics from this point of view here we highlight some of the dynamical systems used to study the mesoscopic physics i.e. the physics in the context of both classical and quantum mechanics of the systems of nanometer dimensions which show mixed (regular and chaotic) behaviour under different experimental conditions.

For such systems, in the quantum domain, one of the important problems is to find a consistent and convergent method by which the energy spectra of classically chaotic Hamiltonian systems could be calculated using the input which is derived exclusively from classical dynamics. For this purpose the use of the Gutzwiller's trace formula [231] for the density of states has been the only theoretical approach to this kind of quantization (called 'quantum chaos') problem for a long time. However, in recent years, several advancements have been made by incorporating the knowledge of both random matrix theory and periodic orbits.

It may be mentioned that the quantum chaos for a system manifests through the statistical properties both of energy (quasi-energy) spectra and eigenfunctions when computed as a function of its classical and quantum (semiclassical) parameters. In this regard while many studies carried out thus far are model based, the general situation turns out to be complicated and no universal properties of quantum chaos can be predicted as such, particularly when the motion in the classical limit is not fully chaotic. As far as the progress on the experimental front is concerned the bulk of available data has been studied from the point of view of both the energy- and time-domains in the theory. While explicitly TD experiments are proliferating and providing further motivation for studying the time-domain theory, other experiments (of course not TD but are best understood in the time-domain because their essential physics is of short duration) are also equally important as far as the understanding of quantum chaos is concerned.

In order to study the influence of quantum effects on chaotic motion of dynamical systems, the model often used is that of kicked-rotator described by the Hamiltonian [232]

$$H = (P_\phi^2 / 2\mathcal{J}) + \epsilon_0 \cos\phi . \delta_T(t), \tag{9.9}$$

where $\delta_T(t) = \sum_{m-\infty}^{\infty} \delta(t - mT)$, is the T-periodic delta function, $\mathcal{J}$ is the moment of inertia and the parameter $\in_0$ is the perturbation strength. Inspite of the restricted form of the external field in (9.9), this model exhibits [232] generic properties of quantum chaos.

Simons et al [233] study the universal correlations of Jastrow type, which describe the response of energy levels of a disordered metallic grain to an arbitrary perturbation. These correlations are shown to be equivalent to TD correlation of a 1D quantum Hamiltonian with inverse-square interaction of Southerland type [233], namely

$$H_S = -\sum_i \frac{\partial^2}{\partial r_i^2} + \beta(\frac{\beta}{2}-1)(\frac{\pi}{N})^2 \sum_{i>j}[\sin^2(\pi(r_i - r_j)/N)]^{-1}, \qquad (9.10)$$

where the spatial coordinates, when expressed in proper units, correspond directly to eigenvalues of the circular ensemble, $\in_i = r_i$, and the coupling of the interaction depends on the Dyson index β which takes the values $\beta = 1, 2$ or 4 corresponding to orthogonal, unitary or symplectic symmetry of the system, respectively. Note that (9.10) describes the motion of particles which are confined by constraining their motion to a ring when the ground states become equivalent to the equilibrium distribution of the circular ensembe of Dyson [234]. In this model, not only the connection between the Hamiltonian (9.10) and the corresponding spectra is much deeper but the time-dependence also appears in a complicated manner. Also, in this case, Wigner-Dyson distribution of eigenvalues (a characteristic feature of quantum chaos) coincides with the probability distribution of the N-particle coordinates of the quantum ground state of (9.10).

There has also been interest in the study of the influence of classical chaos on discrete spectra and their level statistics by using a discrete SE in a quasi periodic potential (known as Harper's equation in the literature [235]) of the type

$$\psi_{n+1} + \psi_{n-1} + \lambda \cos(2\pi n \sigma - \varphi_0)\psi_n = E\psi_n, \qquad (9.11)$$

where ψ_n is the wavefunction at site n and $\lambda = 2$. The dimensionless parameter $\sigma = a^2 eB/(\hbar c)$, gives the number of flux quanta per unit cell of area a^2. In fact eq. (9.11), while used to describe the Bloch electrons in a magnetic field B, is basically a stationary SE, $H\psi = E\psi$, with the Hamiltonian operator of the form

$$H = 2\cos(p) + \lambda \cos(x), \qquad (9.12)$$

where $\lambda = 2$, and $p = -i\hbar\,(\partial/\partial x)$ with an effective $\hbar = 2\pi\sigma$. Geisel et al [236] have further extended these studies to the case of an uncountable (Cantor-type) spectrum. Note that the classical limit of the Hamiltonian (9.12) is integrable and a nonintegrable counterpart of it in the form [237]

$$H = L\cos(p) + K\cos(x).\delta_1(t), \tag{9.13}$$

has also been investigated. Here $\delta_1(t)$ is the periodic delta function of period one and L and K are constants. Clearly, the system (9.13) is also periodic in both x and p. One looks for the solutions of the corresponding TDSE using the Bloch theorem. Since the kicks are periodic in time, the Floquet theorem is applicable and one accordingly writes the quasi-energy eigenstates as $\Psi_\omega(x, t) = \exp(i\theta_x x)\,\psi_\omega(x, t)$, where θ_x is the quasi-momentum and the operator p has eigenvalues of the form $\hbar\,(n + \theta_x)$ with n as an integer. For this system, the time evolution operator U acting on periodic function in x for one period of time is now given by $U = \exp[-i\,(L/\hbar)\cos(\hbar\,(n+\theta_x))]\,.\,\exp[-i\,(K/\hbar)\cos x]$, and determines the quasi-energies ω by $U\,\psi_\omega(x, t) = \exp(i\omega)\,\psi_\omega(x, t)$.

9.3 FUTURE PROSPECTS

At the end of each chapter, no doubt a section on the concluding discussion draws the attention of the reader to the problems left untouched in that chapter yet a central discussion of these different on-going problems at one place is essential. Broadly speaking the future problems to be tackled in relation to the NC potentials can be posed in both classical and quantum contexts.

In the classical context, the problems are open at several stages. First of all, barring a few particular functional forms of the potential (cf. Yoshida's theorem [11]), in general there does not yet exist any criterion or guideline by which one can ensure the existence of the invariant of a system in advance. Although the Painleve′ method provides some such hints but with a limited success. In fact, one has to try the system even for checking the existence of the invariant. Once the existence of the invariant is confirmed, then at the second stage arises the question of its construction, preferably in an exact manner. For this purpose, although various methods have been developed but they still seem inadequate as far as the construction of all (permissible in number) invariants of a system is concerned. From this point of view the situation is presently better for TID systems than that for TD systems.

In general, whether it is a TID or a TD case the search has to be continued either for the integrable systems or for the systems which admit at least a few invariants. Even the availability of a few invariants (few in the sense that the number of invariants is less than the prescribed one according to the criterion for integrability) is helpful as far as the understanding of the physical phenomenon or the simplification of the mathematical computation is concerned. For both TID and TD systems in this book we have restricted ourselves to the construction of exact invariants mainly of polynomial form in momenta and that too upto fourth order in the TID case and upto third order in the TD case. No doubt these polynomial forms of the invariant are easy to handle from the point of view of both physical applications and interpretations, still nonpolynomial forms, if known to exist and become available, could be useful at least for the purposes of the computation of the problem. By constructing the second invariant for some of the TID systems in 2D while we have ensured the integrability of these systems, for TD systems however, it has just not been possible to obtain the second invariant. From this point of view, either the existing methods need updating or else an ansatz for further higher order invariants may be useful. The latter will definitely make the construction rather cumbersome.

With a view to determining the systems admitting invariant(s) of a particular order in momenta the 'potential' equations are set in Chapts. 2 and 3. These potential equations are basically the PDE's and their order is linked with the order of the invariant. We have been able to list only some particular solutions of these potential equations (cf. eqs. (2.19) and (2.20) for the TID case and eqs. (3.45), (3.57) –(3.59), (3.114) and (3.115) for the TD case). It may be of interest to look for other (general) solutions of these potential equations and, if they exist, to obtain accordingly other systems (not listed in the book) admitting invariant of the given order. After a study of coupled TD systems in 2D it appears that a new look is necessary on the Ermakov systems in higher dimensions. In fact, the type of Ermakov systems derived in this book seems to differ from the conventional ones. Their study in polar coordinates can perhaps throw more light in this regard.

In the quantum context, our efforts have been to discuss the methods used to obtain the exact solutions of the stationary state SE and TDSE as well. No doubt the standard-differential equation method or for that matter supersymmetric factorization method are capable of providing exact solutions of the SE with complete spectrum of eigenvalues; however they work well only for a few choices of the potentials. These methods even show limitations when applied to NC and nonseparable potentials in two and higher dimensions, inspite of their successes to the corresponding 1D cases. A variety of potentials can, indeed, be handled exactly by using the eigenfunction ansatz method which somehow provides only a few eigenvalues and that too with certain constraints on the potential parameters. In spite of the fact that these solutions can still be useful for developing higher-order perturbations, the methods which could provide exact solutions of the SE with complete spectrum of eigenvalues and without any constraint on the potential parameters are yet to be discovered. Barring a few other potentials including the exponential ones, by and large polynomial potentials are studied in the literature. In 2D, the cases only upto quartic-type anharmonicity are explored in this book. For the case of higher anharmonicities (sextic or octic) however only approximation methods are available. It could be of interest to apply the eigenfunction-ansatz method to these latter cases and to compare the exact results so obtained with the available approximate ones.

As far as the studies of the SE in literature are concerned, the situation is better for the stationary state case than for the TD case. Except for the one, which is also an extension of the eigenfunction-ansatz method for the TD case in 2D, methods are not yet available to deal with NC TD potentials in 2D. Even this method works well for the case of a TD HO and for TD AH systems this method poses several problems. Regarding other formal methods like time evolution operator or Feynman propagator method the extensions to the 2D case have not yet been carried out.

The last but not the least important future project could be to look for physical interpretations, if possible, of various exact invariants constructed so far for different TID and TD systems.

Finally, it will not be an exaggeration if one says that the study of dynamical systems is basically the study of the on-going problems in various branches of mathematical sciences including physics. Alternatively, the built-in space time structure is the only characteristic feature of all objective sciences which makes them differ from other subjective disciplines (like humanities and social sciences) and the same clearly manifests through the study of dynamical systems. No doubt the scope of study of dynamical systems has been widened after the advancement of computers but even prior to that such studies have been of fundamental nature. It is true that for the purposes of mathematical modeling (and to some extent for mathematical abstraction as well) physicists often deal with idealized situations e.g. for them the gases are perfect, surfaces are smooth, fluids are ideal or nonviscous, conductors are super, forces are not only harmonic or of inverse-square type but are also central in two or higher dimensions etc. In the same spirit their dynamical systems are integrable. Is this not a philosophically much deeper kind of idealized situation? A study of the realistic situation no doubt demands a deviation from these idealized situations but this adds to the complexity not only at the level of methods of understanding of the system but also at the level of the knowledge gained about the system. This is what appears from the present day studies of nonintegrable systems showing chaotic behaviour. Any way, the subject as a whole seems interesting and is rendered easier with the help of available computer codes.

Appendix 1

Field Theoretic Studies in Two Dimensions: Classical Analogue of Yang-Mills Theories

The knowledge of dynamical invariants can also play an important role in the study of classical analogue of Yang-Mills (Y-M) field equations. In view of the fact that solutions to the nonlinear TID Y-M equations are not available, some have been discovered just by accident. None of the systematic methods developed in electrodynamics, such as Fourier decomposition or the Green function method can be applied. The only way to investigate the time-evolution of Y-M fields seems to be rather difficult guessing of examples of solutions. In the case of classical Y-M fields Savvidy et al [193] have shown that the associated uniform magnetic field is unstable. The unstable mode can lower the energy of the ground state. Ambjorn and Olesen [238] have shown that the classical ground state of a Y-M field in the presence of a constant magnetic field is mathematically identical to that of a type II superconductor. This idea has been applied to the magnetic bottling phenomenon of fusion reactions.

A variety of TD solutions of Y-M equations related to the Wu-Yang magnetic monpole have been given by Arodoz [239]. It has been shown by Nikolaevskii and Shur [240] that there is no first integral for a particular case of the classical Y-M equations and the original system of equations as such does not have a complete set of integrals. As a matter of fact the question of integrability of the classical Y-M equations is exceedingly important for both classical and quantum field theories.

The Y-M system is described by the Lagrangian density (cf. Sect. 7.2.6)

$$\mathcal{L} = -\frac{1}{4} F^a_{\mu\upsilon}\, F^a_{\mu\upsilon} \,, \tag{A1.1}$$

where $F^a_{\mu\upsilon} = \partial_\mu A^a_\upsilon - \partial_\upsilon A^a_\mu + g \in^{abc} A^b_\mu A^c_\upsilon$, and the corresponding field equation is given by

$$\partial_\mu F^a_{\mu\upsilon} + g \in^{abc} A^b_\mu F^c_{\mu\upsilon} = 0. \tag{A1.2}$$

Here $\mu, \upsilon = 0, 1, 2, 3, ; a, b, c, = 1,2,3;$ A^a_μ are the elements of an arbitrary Lie algebra and g is the metric tensor. The solutions of (A1.2) in which we are interested, correspond to the SU (2) case and to the gauge

$$A^a_0 = 0, \quad A^a_i = A^a_i(t). \tag{A1.3}$$

In this gauge, the field eq. (A1.2) gives the Gauss law in the form

$$g \in^{abc} A^b_\mu F^c_{\mu 0} = 0. \tag{A1.4}$$

On the other hand, the definition of $F^a_{\mu\upsilon}$ using (A1.3) gives

$$\left.\begin{aligned} F^c_{io} &= \partial_i A^c_0 - \partial_0 A^c_i + g \in^{bac} A^a_i A^b_0 = -\partial_0 A^c_i, \\ F^c_{oi} &= \partial_o A^c_i; \qquad F^a_{ij} = g \in^{abc} A^a_i A^c_j, \end{aligned}\right\} \tag{A1.5}$$

which, in turn, from (A1.4) provide

$$g \in^{abc} A^b_i (-\partial_o A^c_i) = 0,$$

or

$$A^b_i \partial_o A^c_i - A^c_i \partial_o A^b_i = 0. \tag{A1.6}$$

Substituting $\upsilon = i$ and $\mu = j$ in (A1.2) and after using (A1.5) one obtains the field eq. (A1.2) in the form

$$\partial^2_o A^a_i + g^2 (A^a_i A^b_j - A^a_j A^b_i) A^b_j = 0. \tag{A1.7}$$

One can satisfy the Gauss law (A1.6) by a 9-parameter form [241]

$$A^a_i = O^a_i f^a(t), \; (a \text{ not summed}), \tag{A1.8}$$

where O^a_i's are constant orthogonal matrices obeying

$$O^a_i O^b_i = (1/g^2)\,\delta^{ab}. \tag{A1.9}$$

Using (A1.8), eq. (A1.7) reduces to the form (cf. eq. (7.49′))

$$\partial^2_o f^a + g^2 f^a (O^b_j f^b O^b_j f^b - \frac{1}{g^2} f^a f^b) = 0$$

or
$$\ddot{f}^a + \sum_{b \neq a} (f^b)^2 f^a = 0. \tag{A1.10}$$

This system (A1.10), expressible by an effective Lagrangian

$$L = \frac{1}{2} \sum_a (\dot{f}^a)^2 - \frac{1}{4} \sum_a \sum_{b \neq a} (f^a)^2 (f^b)^2, \tag{A1.11}$$

can be shown to have positive-definite conserved energy and has been the subject of study by several authors [193, 194] for various special cases. In fact, depending upon the symmetry group a variety of classical dynamical systems out of the general structure (A1.10) or (A1.11) in conjunction with (A1.8) are investigated with reference to the integrability of the resultant system. We briefly survey some of these studies here:

With a slightly modified definition of the matrix O_i^a in (A1.8) and for the substitution $A_1^1 = x_1$, $A_2^2 = x_2$ and with all other components of A_i^a zero, the Hamiltonian structure studied by Savvidy and his coworkers [193, 241] corresponds to

$$H = \frac{1}{2}(p_1^2 + p_2^2) + \frac{1}{4} g^2 \eta^2 (x_1^2 + x_2^2) + \frac{1}{2} g^2 (x_1 x_2)^2. \tag{A1.12}$$

Similarly, corresponding to the SO (3) ⊗ SO (3) symmetry which preserves the moments (corresponding to the coupling eq. (A1.4))

$$n^a = \epsilon^{abc} A_i^b \dot{A}_i^c \; ; \; m_i = \epsilon_{ijk} A_j^a \dot{A}_k^a ,$$

another form investigated by them is of somewhat complicated nature. The same is described by the Hamiltonian (see e.g. Savvidy in Ref. [193])

$$H_{YM} = \frac{1}{2}(p_1^2 + p_2^2 + p_3^2) + \frac{1}{2} g^2 (x_1^2 x_2^2 + x_2^2 x_3^2 + x_3^2 x_1^2) \; T_{YM}, \tag{A1.13}$$

where
$$T_{YM} = \frac{1}{2} \sum_{a,i} (I_i \, \Omega_i^2 + I_a \, \omega_a^2 - 2 I_{ai} \, \Omega_i \, \omega_a)$$

with $I_1 = x_2^2 + x_3^2$; $I_2 = x_3^2 + x_1^2$; $I_3 = x_1^2 + x_2^2$; $I_{11} = 2x_2 \, x_3$; $I_{22} = 2x_1 \, x_3$, $I_{33} = 2x_1 \, x_2$; and $I_{ai} = 0$ (for $i \neq a$). For the SU(2) case and under the adiabatic approximation regarding the time-dependence of f^a's in A1.10), Chang [193] has analysed the iterative maps associated with eqs. (A1.10) for $a = 1,2$. In fact, for this case by defining $f^1 = x_1$, $f^2 = x_2$, the system of eqs. (A1.10) reduces to

$$\ddot{x}_1 = -x_1 \, x_2^2 \; ; \; \ddot{x}_2 = -x_2 \, x_1^2, \tag{A1.14}$$

which can be derived from the Hamiltonian

$$H_1 = \frac{1}{2}(p_1^2 + p_2^2) + \frac{1}{2} x_1^2 x_2^2. \tag{A1.15}$$

It is conjectured in the literature [193, 242] that this system is strongly chaotic and nonintegrable. Numerical investigations by Carnegie and Percival [242] suggest that no regular regions of motion exist on the surface of section and the motion is always irregular, except for a measure zero set of unstable periodic orbits. On the other hand, the results of Dahlqvist and Russberg [243] indicate that there exists at least one family of stable orbits for the system (A1.15). These latter results are based on the studies of the one-parameter potential, $V(x_1, x_2) = \frac{1}{2}(x_1^2 x_2^2)^{1/\bar{a}}$, from where the system (A1.15) is obtained for $\bar{a} = 1$.

Without actually going to the ansatz (A1.8), Villarroel [194] directly analyses the system (A1.7) with the conditions (A1.4) and $A_i^3 = A_3^a = 0$, for $i = 1, 2, 3$; $a = 1, 2, 3$. As a result he investigates a system with four degrees of freedom described by the set of equations ($A_1^1 \equiv x_1, A_1^2 \equiv x_2, A_2^1 \equiv x_3, A_2^2 \equiv x_4$)

$$\begin{aligned} \ddot{x}_1 + g^2 x_4(x_1x_4 - x_2x_3) &= 0, \\ \ddot{x}_2 - g^2 x_3(x_1x_4 - x_2x_3) &= 0, \\ \ddot{x}_3 - g^2 x_2(x_1x_4 - x_2x_3) &= 0, \\ \ddot{x}_4 + g^2 x_1(x_1x_4 - x_2x_3) &= 0, \end{aligned} \tag{A1.16}$$

alongwith the Gauss law (A1.4) in the form $g^2(x_2\dot{x}_1 - x_1\dot{x}_2 + x_4\dot{x}_3 - x_3\dot{x}_4) = 0$. Note that this system can again be derived from the Hamiltonian

$$H_2 = \frac{1}{2}(p_1^2 + p_2^2 + p_3^2 + p_4^2) + \frac{1}{2} g^2 (x_1x_4 - x_2x_3)^2. \tag{A1.17}$$

The integrability of the systems (A1.15) and (A1.17) is analysed by Ichtiaroglou [194] in view of the rigorous results of Yoshida [11].

Thus, we find that the Y-M classical mechanics offers a treasure of dynamical systems and here we have presented only a few of them. Unfortunately, the integrability of these subsystems is not yet established. While the study of these systems appears to be only of academic interest at present, some of them can definitely be a part of the usual classical mechanics. Not only this Savvidy [193] also cites a subsystem of Y-M theory (if one ignores the problem of quantization of nonintegrable systems) which is the quantum analogue of Y-M mechanics. Under highly restricted conditions the SE for such a system (which, in fact, is a special case of (A1.13)) can be written as

$$-\frac{1}{2}\nabla^2\psi + [\frac{1}{2}g^2(x_1^2x_2^2 + x_2^2x_3^2 + x_3^2x_1^2) + U_{eff} - \mathcal{E}\,]\psi = 0. \qquad \text{(A1.18)}$$

Since the corresponding system is nonintegrable, the SE (A1.18) can not be solved by the method of separation of variables.

Appendix 2

Classical and Quantum Integrability

Inspite of the fact that nonintegrable systems outnumber the integrable ones, the concept of integrability, if holds for a system, definitely enlightens some of its distinct features in the same way as the symmetries, if found to exist for a physical system, reveal a lot about its systematic properties. The concept of integrability has been used differently (but in a related manner) in the context of classical mechanics, quantum mechanics and field theory. Also, this concept is used differently for finite- and infinite-dimensional systems. However, here our discussion of this concept will be restricted only to finite-dimensional systems in the context of classical and quantum mechanics alone.

A classical Hamiltonian system in D dimensions, as mentioned in Chap. 1, is said to be classically integrable if there exist (D-1) independent, well defined, global functions, whose Poisson brackets with each other and with the Hamiltonian vanish. On the other hand, a quantum mechanical Hamiltonian of D dimensions is said to be quantum integrable if there are (D-1) independent, well defined global operators, which commute with each other and with the Hamiltonian. The comparison between classical and quantum integrability becomes easier if one can represent the quantum operators by c-number functions. Since classical and quantum mechanics are not algebraically isomorphic, differences are bound to appear even if one uses the c-number functions in both. As a matter of fact the commutator will become the Moyal bracket [245] which in turn reduces to the Poisson bracket only when $\hbar \to 0$.

In this connection several correspondence rules are discussed in the literature [244]. In fact, to construct quantum mechanics out of classical mechanics different correspondence rules would give different operators for the same function of classical mechanics. However, when the operator is fixed these different correspondence rules, characterized by a function of 2n variables $\mathcal{F}(x, y)$, will then give different c-number representations. The formal integral used to demonstrate these correspondence rules between the c-number function $A(p,q)$ and the quantum operator $\hat{\mathcal{A}}(\hat{p},\hat{q})$ is given by [244, 249].

$$\hat{\mathcal{A}}(\hat{p},\hat{q}) = \int d^nq\, d^np\, d^nx\, d^ny\, (2p\hbar)^{-2n}\, \mathcal{F}(x,y)\, A_{\mathcal{F}}(p,q).$$

$$\exp\,[\, i\{x.(\hat{p}-p) + y.(\hat{q}-q)\}\,/\,\hbar], \qquad \text{(A2.1)}$$

where the label $\mathcal{F}$ to A indicates that the operator $\hat{\mathcal{A}}$ is fixed and $A_{\mathcal{F}}(p, q)$ will be different for different $\mathcal{F}(x, y)$. For all practical purposes while $\mathcal{F}(x, y) = F(i\,(x\,.\,y)/\hbar)$ with $F(0) = 1$ is assumed, Weyl and standard ordering rules corresponding to $\mathcal{F}_w(x, y) = 1$ and $\mathcal{F}_s(x, y) = \exp[-i\,(x.\,y)/2\hbar)]$ are found to be more convenient. For details we refer to the work of Hietarinta [249]. Once the correspondence (A2.1) is given, the relationship between the operators $\hat{\mathcal{A}}, \hat{\mathcal{B}}, \hat{\mathcal{C}}$ in the form $[\hat{\mathcal{A}},\hat{\mathcal{B}}] = i\hbar\,\hat{\mathcal{C}}$, can be translated into the corresponding c-number representations A, B, C for Weyl ordering rule $(F \equiv 1)$ as [245, 249]

$$C(p,q) \equiv \{A,B\}_{MB} = \frac{2}{\hbar} A \sin\left(\frac{1}{2}\hbar\,\overset{\leftrightarrow}{\Lambda}\right) B$$

$$= A\overset{\leftrightarrow}{\Lambda}B - \frac{1}{24}\hbar^2\, A\overset{\leftrightarrow 3}{\Lambda}B + \frac{1}{1920}\hbar^4\, A\overset{\leftrightarrow 5}{\Lambda}B + \ldots\,. \qquad \text{(A2.2)}$$

Here $\{A, B\}_{\mathrm{MB}}$ denotes what is called Moyal bracket and $\overset{\leftrightarrow}{\Lambda}$ for a D-dimensional system is given by

$$\overset{\leftrightarrow}{\Lambda} = \sum_{i=1}^{D} \left(\frac{\overleftarrow{\partial}}{\partial q_i}\frac{\overrightarrow{\partial}}{\partial p_i} - \frac{\overleftarrow{\partial}}{\partial p_i}\frac{\overrightarrow{\partial}}{\partial q_i} \right). \qquad \text{(A2.3)}$$

Certain remarks and observations regarding the integrability of a classical dynamical system are made in Sect. 1.4. While the notion of this concept is more transparent at the classical level in the spirit of Liouville's theorem [9], at the levels of quantum mechanics and field theory, however, it requires [246, 247] further investigations. Against the belief [248] that the classical integrability of a system implies its quantum integrability in a trivial manner, investigations have shown [249, 250] that this indeed is not the case. In fact, in order to obtain the quantum invariant of a system from the corresponding classical one, the quantum corrections, arising from the terms involving $\hbar$ in the expansion of the sine function in (A2.2), need to be incorporated. It can be seen from (A2.2) that in the lowest order in the powers of $\hbar$, $\{A,B\}_{\rm MB} = A \overleftrightarrow{\Lambda} B$ which is just the Poisson bracket $[A,B]_{\rm PB}$. Such quantum corrections to the second constant of motion have been computed [249, 250] for a number of TID systems in 2D upto $O(\hbar^2)$. It may be mentioned that if the second constant of motion is at most of second order in momenta (like the Hamiltonian of the system), then the Moyal bracket (A2.2) exactly reduces to the Poisson bracket. In this case classical and quantum invariants turn out to be identical. For cubic and higher order invariants however the quantum corrections arise. In fact, it is only after these corrections a classical invariant becomes quantum invariant in the spirit of Moyal bracket. While these corrections are obtained for TID systems for different orders of the polynomial (in momenta) invariants, they are not yet derived for the TD systems. Here, following the work of Hietarinta [249, 250], we briefly present the quantum analogue of the classical (second) invariant for a number of TID systems upto $O(\hbar^2)$ alongwith the method adopted to obtain these results.

Cubic and quartic invariants: Recall that for the classical case the invariant of a system described by the Hamiltonian H is obtained from $(dI/dt) = [I,H]_{\rm PB} = 0$, whereas for the quantum case it is now obtained from

$$(dI/dt) = \{I,H\}_{\rm MB} = 0. \tag{A2.4}$$

The use of the Moyal bracket from (A2.2) here in the latter equation and its expansion in the powers of $\hbar$ in the process of construction of the quantum invariant for a 2D system will affect the equations of the type (2.15) and (2.16). Thus, for the corresponding Cartesian case and for the existence of the quantum invariant of cubic order up to $O(\hbar^2)$, eq. (2.15*i*) will now be modified to the form

$$a_1\partial_{x_1}V + a_2\partial_{x_2}V - \frac{1}{24}\hbar^2[a_{111}\partial^3_{x_1}V + a_{112}\partial^2_{x_1}\partial_{x_2}V + a_{122}\partial_{x_1}\partial^2_{x_2}V + \\ + a_{222}\partial^3_{x_2}V] = 0. \tag{A2.5}$$

In view of (A2.5) one can state a general result as follows: If the Hamiltonian

$$H = \frac{1}{2}(p_1^2 + p_2^2) + V(x_1,x_2)\,, \tag{A2.6}$$

is classically integrable and has the second invariant of the form (cf. Chap. 2)

$$I_3 = a_i\,\xi_i + \frac{1}{6}\,a_{ijk}\,\xi_i\,\xi_j\,\xi_k\,, \quad (\xi_1 = p_1,\ \xi_2 = p_2), \quad \text{(A2.7)}$$

then the corresponding quantum Hamiltonian is also integrable upto $O(\hbar^2)$ and has the second invariant $\hat{I}_3$, provided

$$a_{111}\partial^3_{x_1}V + a_{112}\partial^2_{x_1}\partial_{x_2}V + a_{122}\partial_{x_1}\partial^2_{x_2}V + a_{222}\partial^3_{x_2}V = 0. \quad \text{(A2.8)}$$

A similar result can also be derived for the invariants of quartic order in momenta, namely for (cf. Chap. 2)

$$I_4 = a_o + \frac{1}{2}a_{ij}\,\xi_i\xi_j + \frac{1}{24}a_{ijkl}\,\xi_i\xi_j\xi_k\xi_l\,, \qquad (\xi_1 = p_1, \xi_2 = p_2). \quad \text{(A2.9)}$$

In this context, for the corresponding Cartesian case the equations of the type (2.16k) and (2.16ℓ) will be modified to the form [249]

$$\partial_{x_1}a_o = a_{11}\partial_{x_1}V + \frac{1}{2}\,a_{12}\partial_{x_2}V - \frac{\hbar^2}{96}\,[4a_{1111}\partial^3_{x_1}V + 3a_{1112}\partial^2_{x_1}\partial_{x_2}V + + 2a_{1122}\partial_{x_1}\partial^2_{x_2}V + a_{1222}\partial^3_{x_2}V\,], \quad \text{(A2.10)}$$

$$\partial_{x_2}a_o = \frac{1}{2}\,a_{12}\partial_{x_1}V + a_{22}\partial_{x_2}V - \frac{\hbar^2}{96}\,[\,a_{1112}\partial^3_{x_1}V + 2a_{1122}\partial^2_{x_1}\partial_{x_2}V + + 3a_{1222}\partial_{x_1}\partial^2_{x_2}V + 4a_{2222}\partial^3_{x_2}V\,]. \quad \text{(A2.11)}$$

Further a general result, as for the cubic case, can be stated as follows: If the Hamiltonian (A2.6) is classically integrable and has second invariant of the form (A2.9) then the corresponding quantum Hamiltonian is also integrable upto $O(\hbar^2)$ and has the second invariant $\hat{I}_4$, provided the coefficients of $\hbar^2$ in eqs. (A2.10) and (A2.11) vanish.

It may be mentioned that the conditions (A2.8), (A2.10) and (A2.11) are sufficient but not always necessary for a system to be both classically and quantally integrable. In fact, in some cases the quantum corrections also appear with the coefficient functions a_o, a_i, a_{ij}, a_{ijk} and a_{ijkl} in (A2.7) and (A2.9). This very much depends on the nature of the potential term in (A2.6). We now discuss the integrability of some typical cases [249].

(*i*) The Toda-type potential,

$$V(x_1,x_2) = \exp(\sqrt{3}x+y) + \exp(-\sqrt{3}x+y) + \exp(-2y),$$

admitting a cubic invariant is found to be both classically and quantally integrable.

(*ii*) The Holt potential [13],

$$V(x_1,x_2) = \frac{3}{4}x_1^{4/3} + x_2^2\, x_1^{-2/3},$$

admitting a cubic invariant, is found only to be classically integrable.

(*iii*) The Henon-Heiles type potential

$$V(x_1,x_2) \doteq \frac{16}{3}x_2^3 + x_1^2\, x_2,$$

admitting a quartic invariant,

$$I_4 = p_1^4 + 4x_2\, x_1^2\, p_1^2 - \frac{4}{3}x_1^3\, p_1\, p_2 - \frac{4}{3}x_2^2\, x_1^4 - \frac{2}{9}x_1^6,$$

turns out to be classically as well as quantally integrable. For the classical and quantum integrability of several other systems including the variants of the above cases, we refer to the work of Hietarinta [249].

Deformations of the potential: Note that some of the potentials (like the Holt-potential) discussed above are found to be classically integrable but not quantally. Here we discuss the possibility that a classically integrable potential can be made quantum integrable when the potential is deformed by a term of $O(\hbar^2)$. For this purpose, write (A2.6) as

$$H = \frac{1}{2}(p_1^2 + p_2^2) + V(x_1, x_2) + \hbar^2\, \Delta(x_1,x_2), \qquad \text{(A2.12)}$$

and then construct the second invariant order-by-order in $\hbar^2$. For the case of cubic invariants one can make the following ansatz for the connection between classical and quantum results

$$I_3^Q = I_3^c + \hbar^2\, (p_1 A_1(x_1,x_2) + p_2 A_2(x_1,x_2)), \qquad \text{(A2.13)}$$

where the functions A_1 and A_2 can be determined in terms of $\Delta(x_1, x_2)$ from

$$\partial_{x_1} A_1 = \frac{1}{2}a_{111}\partial_{x_1}\Delta + \frac{1}{6}a_{112}\partial_{x_2}\Delta;\, \partial_{x_2} A_2 = \frac{1}{6}a_{122}\partial_{x_1}\Delta + \frac{1}{2}a_{222}\partial_{x_2}\Delta;$$

$$\partial_{x_2} A_1 + \partial_{x_1} A_2 = \frac{1}{3} a_{112} \partial_{x_1} \Delta + \frac{1}{3} a_{122} \partial_{x_2} \Delta .$$

with a_{ijk} determined already as before.

For example, for the Holt Hamiltonian

$$H = \frac{1}{2}(p_1^2 + p_2^2) + \frac{3}{4} x_1^{4/3} + x_2^2 x_1^{-2/3} + \delta x_1^{-2/3},$$

which admits the classical invariant,

$$I_3 = p_1^3 + \frac{3}{2} p_2 p_1^2 + (-\frac{9}{2} x_1^{4/3} + 3x_1^{-2/3} x_2^2 + 3\delta x_1^{-2/3}) p_2 + 9x_1^{1/3} x_2 p_1,$$

$\Delta_Q H$ and $\Delta_Q I_3$ turn out to be [249]

$$\Delta_Q H \equiv \hbar^2 \Delta(x_1, x_2) = -\frac{5}{72} \hbar^2 x_1^{-2}$$

$$\Delta_Q I_3 = -\frac{5}{24} \hbar^2 x_1^{-2} p_2.$$

Similarly, for the Hamiltonian [32]

$$H = \frac{1}{2} (p_1^2 + p_2^2) + (x_1 x_2)^{-2/3},$$

admitting the second invariant,

$$I_3 = p_1 p_2 (x_1 p_2 - x_2 p_1) + 2(x_1 x_2)^{-2/3} (p_1 x_1 - p_2 x_2),$$

the quantum deformations are found to be [249, 251]

$$\Delta_Q H \equiv -\frac{5}{72} \hbar^2 (1/x_1^2 + 1/x_2^2);$$

$$\Delta_Q I_3 = -\frac{5}{36} \hbar^2 [(p_1 x_1 / x_2^2) - (p_2 x_2 / x_1^2)].$$

For the Hamiltonian (A2.6) which admits a p-quartic invariant of the form

$$\tilde{I}_4 = p_1^4 + A(x_1, x_2) p_1^2 + B(x_1, x_2) p_1 p_2 + C(x_1, x_2) p_2^2 + D(x_1, x_2), \qquad \text{(A2.14)}$$

Hietarinta and Grammaticos [250] have tried to obtain a general result as far as its classical and quantal integrability are concerned. In fact, following the work of Gammaticos et al. [35], they have shown that when $V(x_1,x_2)$ in (A2.6) has the form (ν and α's are constants)

$$V(x_1,x_2) = \nu\, x_2^4 + x_2^2\,[f_2''(x_1) + \alpha] + f_o(x_1), \qquad \text{(A2.15)}$$

then the corresponding invariant can be expressed by (A2.14) with

$$A = 4[x_2^2\, f_2''(x_1) + f_0\,(x_1)];\; B = -8\,x_2\, f_2'(x_1);\; C = 8\, f_2(x_1),$$

and the expression for D compatible with $\{H,\tilde{I}_4\}_{MB} = 0 = [H,\tilde{I}_4]_{PB} - 3\hbar^2\, p_1\,(\partial^3 V/\partial x_1^3) = 0$ turns out to be

$$D = -2\,x_2^4[f_2'''f_2 - 8\,\nu\, f_2] + 4\,x_2^2[4f_2''f_2 - f_2'\,f_0' + 4f_2\alpha] - \hbar^2 f_0' + 4f_0^2,$$

where $f_0(x_1)$ is expressible in terms of f_2 as

$$f_0 = -\frac{5}{8}\,\hbar^2\,(f_2''/\,f_2')^2 + 2f_2 + (4\alpha\, f_2^2 - 2\alpha_1\, f_2 + \alpha_2)\,/\,f_2'^2\,,$$

and f_2 must be a solution of

$$f_2''''f_2' + 5f_2'''f_2'' - 24\,\nu\, f_2' = 0,$$

where various symbols have their usual meanings [250]. It may be noted that as the limiting cases of (A2.15) several interesting well known systems like Henon-Heiles or Holt potentials can be obtained.

In spite of all these studies the existence of additive terms in the Hamiltonian in order to make the system also quantum integrable, remains a puzzling phenomenon. As a matter of fact something more than just being ad hoc corrections which restore quantal integrability is required, in the sense of the possible origin of these terms and their physical meanings as well. Not only this, the corrections arising from the terms of $O(\hbar^4)$ and higher, if found nonzero then can provide further insight into the quantal integrability of the system. Studies have also been carried out [147, 251] to find a connection between classical and quantum invariants of a system from the point of view of its symmetry considerations. Such works however have limited scope as far as the study of noncentral potentials are concerned.

Next we consider a system in 3D i.e. the case of the Kowalevskaya top and discuss its classical and quantal integrability.

Integrability of Kowalevskaya top: The case of a heavy top rotating about a fixed point was investigated by the Russian lady mathematician

S. Kowalevskaya [252] as early as 1890 and now classical mechanics of this system is given in several books (see, for example, Ref. [141] Chap. 8). The quantum integrability of this system has also been investigated by Ramani et al. [253].

It is well known that the equations of motion of an asymmetric top around a fixed point under the influence of its own weight can be written as [141, 252, 253]

$$A\,(d\Omega_1/dt) = (B-C)\,\Omega_2\Omega_3 + z_o\beta - y_o\gamma\,, \tag{A2.16a}$$

$$B\,(d\Omega_2/dt) = (C-A)\,\Omega_3\Omega_1 + x_o\gamma - z_o\alpha\,, \tag{A2.16b}$$

$$C\,(d\Omega_3/dt) = (A-B)\,\Omega_1\Omega_2 + y_o\alpha - x_o\beta\,, \tag{A2.16c}$$

$$(d\alpha/dt) = \Omega_3\beta - \Omega_2\gamma;\ (d\beta/dt) = \Omega_1\gamma - \Omega_3\alpha;\ (d\gamma/dt) = \Omega_2\alpha - \Omega_1\beta, \qquad \text{(A2.16d, e, f)}$$

where $\Omega_1, \Omega_2, \Omega_3$ are the components of the angular velocity; α, β, γ are the direction cosines $(\alpha^2 + \beta^2 + \gamma^2 = 1)$; A, B, C are the moments of inertia of the top, and the constants x_o, y_o, z_o, are related to the position of the fixed point with respect to the centre of mass of the top. This system, while has an inherent Hamiltonian structure, admits two integrals of motion, namely the energy

$$E = \frac{1}{2}(A\Omega_1^2 + B\Omega_2^2 + C\Omega_3^2) + x_0\alpha + y_0\beta + z_o\gamma\,, \tag{A2.17}$$

and the vertical component of the angular momentum vector

$$M_Z = A\Omega_1\,\alpha + B\Omega_2\,\beta + C\,\Omega_3\gamma\,. \tag{A2.18}$$

In general, the system (A2.16) is not integrable as no third invariant exists. However, in particular cases the third invariant was constructed by Euler, Lagrange and Kowalevskaya.

For a free top, when the fixed point coincides with its centre of mass (i.e. $x_o = y_o = z_o$), Euler obtained the third invariant as $M^2 = A^2\Omega_1^2 + B^2\Omega_2^2 + C^2\Omega_3^2$, which represents the square of the angular momentum with respect to the body fixed axes. The case of a symmetric top corresponding to $A = B$ and $x_o = y_o = 0$, was investigated by Lagrange and obtained the third invariant as the third projection of the angular momentum, namely $M_3 = C\,\Omega_3$. Kowalevskaya [252] analyzed the problem from the point of view of the analytic structure of the solutions in the complex time plane (cf. Sect. 2.3.2). Besides recovering the cases of Euler and Lagrange in this approach, she obtained the third invariant for the case (known as the Kowalevskaya top) $A = B = 2C$ and $z_o = 0$, as [253]

$$K = (\Omega_1^2 - \Omega_2^2 - x_0\alpha)^2 + (2\Omega_1\Omega_2 - x_0\beta)^2, \qquad \text{(A2.19)}$$

where y_o is set equal to zero without loss of generality.

By writing (A2.17), (A2.18) and (A2.19) in terms of Euler angles θ, ϕ, ψ within the framework of the Hamiltonian formulation of the problem as

$$H = \frac{1}{4}\operatorname{cosec}^2\theta\,(p_\phi^2 + p_\psi^2 - 2p_\phi p_\psi \cos\theta) + \frac{1}{4}\,p_\psi^2 + \frac{1}{4}p_\theta^2 + x_0 \sin\psi \sin\theta, \qquad \text{(A2.20a)}$$

$$M = p_\phi, \qquad \text{(A2.20b)}$$

$$K = \frac{1}{16}\operatorname{cosec}^2\theta\,[\,p_\theta^2 + (p_\phi - p_\psi \cos\theta)^2] + x_0^2 \sin^2\theta - \frac{1}{2}\,x_0$$

$$[\{\operatorname{cosec}^2\theta\,(p_\phi - p_\psi \cos\theta)^2 - p_\theta^2\}\sin\psi \sin\theta + 2p_\theta\,(p_\phi - p_\psi \cos\theta)\cos\psi], \qquad \text{(A2.20c)}$$

where $C = 1$ and $y_o = 0$ is used, Ramani et al [253] have investigated the quantum integrabily of this system up to $O(\hbar^4)$. They use a computer code for their analytical work and write the c-number version of the operators $\hat{H}$ and $\hat{K}$ as

$$H' = H_0 + h^2 H_2\,, \quad K' = K_0 + h^2 K_2 + h^4 K_4\,, \qquad \text{(A2.21)}$$

and then look for the nonvanishing terms in the Moyal bracket $H', K'\}_{MB}$ of orders h^2 and h^4. It is found that for

$$H_2 = -\frac{1}{4}\operatorname{cosec}^2\theta,$$

$$K_2 = -\frac{7}{8}\,(p_\phi - p_\psi \cos\theta)^2 \operatorname{cosec}^2\theta + \frac{1}{4}\,(2p_\phi^2 - p_\phi p_\psi \cos\theta$$

$$-p_\psi^2 - \frac{1}{2}\,p_\theta^2)\operatorname{cosec}^2\theta,$$

$$K_4 = \frac{13}{16}\operatorname{cosec}^4\theta - \frac{1}{2}\operatorname{cosec}^2\theta,$$

the corrective terms in (A2.21) ensure the quantum integrability of the system (A2.16) in the Weyl representation.

Appendix 3

Group Theoretical Methods and Constants of Motion

Inspite of the fact that a detailed discussion of the role of *group theoretical methods* in both classical and quantum contexts is avoided all throughout the book, some flavour of these elegant methods can be noticed in Sects. 2.3.3 and 6.1.1 (in connection with Lie symmetries and associated integrals of motion), in Sect. 3.5.1 (in connection with generalized Ermakov systems) and in Sects. 4.2.2. (c) and 4.7 (in connection with exact solutions of the SE). Here mainly for the sake of completeness, following the works of Fokas [254] and Fokas and Lagerstrom [32] a brief discussion of group theoretical aspects of constants of motion, particularly in the classical context, is presented.

The use of group theoretical methods for understanding and solving the problems of both classical and quantum mechanics has been well known [16, 49, 147, 218]. Inspite of this there exist several fundamental questions even today and some of them can be posed as follows: In the classical context (i) while the constants of motion, linear in momenta, are related to Lie (point) groups of the Hamilton-Jacobi equation, in general there is no way of using Lie theory to account for the existence of conserved quantities which are nonlinear in momenta; (ii) inspite of difficulties in group theoretical characterization of all separable solutions of the Hamilton-Jacobi equation, the use of Lie theory suggests the characterization only of some of the separable solutions. In the quantum context, on the other hand, Noether's theorem [42] while connects conservation laws and invariant properties of a given system and is based on the fact that the system possesses a Lagrangian, it cannot guarantee that every conservation law can be derived from Lie groups. (For example, the quantum mechanical analogue of Runge-Lenz vector for the hydrogen atom is not the consequence of Lie groups.) For this purpose, although the symmetries of non-geometric nature (dynamical symmetries) were introduced [147] (cf. Sect. 4.7) nevertheless the search for exact group nature of these symmetries from the point of view of generalization of Lie theory still remains a problem.

Ibragimov and Anderson [255] tried to resolve some of these problems by introducing what are known as Lie-Backlund groups which were found useful for the purposes of generalizations of the Noether's theorem. In this regard the dynamical symmetries which were initially related to the problem of separation of variables of the SE, could as well be the manifestations of "higher" symmetries admitted by the potential (cf. Sect. 4.7). The role of these higher (Lie-Backlund) symmetries, if properly explored in classical mechanics can also provide a better understanding of the corresponding quantum problem. Indeed it turns out that the constants of motion, linear in momenta, correspond to Lie groups while those which are nonlinear in momenta correspond, in general, to Lie-Backlund groups. Moreover, the existence of Lie-Backlund groups in classical mechanics not only leads to conserved quantities of Hamilton's equations and separable solutions of Hamilton-Jacobi equation but also to Lie groups of Hamilton's equations. For details we refer to the relevant literature [254-256]. Here, we just outline the general procedure followed for the construction of conservation laws from symmetries of a differential equation using Noether's theorem. Before bringing out the essence of this procedure it is worth recalling some of the salient points.

As a matter of fact each constant of motion, if known for a system restricts its motion and the same can be used to reduce by one the number of degrees of freedom of the system. The search for the conservation laws (which directly provide the constants of motion) for a system is generally a first step towards finding the solution of the problem. More the conservation laws one finds the closer one gets to the complete solution. While for the Lagrangian formulation of the problem Noether's theorem offers these conservation laws, it is, in general, a difficult task to construct these conservation laws for a given system. In fact, for every infinitesimal transformation admitted by the action integral of a Lagrangian system, one can construct a conservation law. The important point is that all such transformations can be found by examining the invariance properties of the corresponding Lagrange's equations of motion which are essentially the differential equations arising from a variational problem of the action integral. Again a number of problems can be expressed in this type of variational formulation.

Consider a physical system which has independent variables t (time) and $x \equiv (x_1, x_2, \dots, x_n) \in \mathcal{R}^N$, and dependent variables described by a state function $u(t, x_1, \dots\dots, x_n)$ with components $u \equiv (u^1, u^2, \dots, u^M) \in \mathcal{R}^M$.

The partial derivatives of u for every $k = 1, 2$ .. are labelled as $u^{\alpha}_{i_1, i_2, \dots, i_k}$, ($\alpha=1, \dots, M$; $i_1, i_2 \dots i_k = 1, \dots, N$) which is symmetric in their lower indices. For a one-parameter group G, the point transformations can be expressed as

$$
\text{G:}\qquad
\begin{aligned}
x'_i &= g_i(x, u, \underset{1}{u}, \underset{2}{u}, \dots\dots, \underset{k}{u}, t),\\
u'^{\alpha} &= \phi^{\alpha}(x, u, \underset{1}{u}, \underset{2}{u}, \dots\dots, \underset{k}{u}, t),\\
u'^{\alpha}_i &= \psi^{\alpha}_i(x, u, \underset{1}{u}, \underset{2}{u}, \dots\dots, \underset{k}{u}, t),\\
&\dots\dots\dots\dots\\
&\dots\dots\dots\dots
\end{aligned}
\qquad \text{(A3.1)}
$$

where $\underset{1}{u}, \underset{2}{u}, \dots, \underset{k}{u}$ denote the derivatives of u and the numerals below u's indicate the number of lower indices to the corresponding u. Now, for a given (Lagrangian) function L (x_1, u, $\underset{1}{u}, \underset{2}{u}, \dots, \underset{k}{u}$) defined in the domain Ω in the space $x \equiv (x_1, \dots x_n)$, one can find functions $u(x)$ which correspond to extrema of the (action) integral

$$
J(u) = \int_{\Omega} L\,(x, u, \underset{1}{u}, \underset{2}{u}, \dots\dots, \underset{k}{u})\, dx. \qquad \text{(A3.2)}
$$

Here $u(x)$ describes the state of the system and is usually subject to a set of conditions prescribed on the boundary $\delta\Omega$ of the domain Ω. If $u(x)$ is an extremum of (A3.2), then any infinitesimal change $u(x) \rightarrow u(x) + \in v(x)$ which does not alter the boundary condition will have no effect on $J(u)$ at least to order $O(\in)$.

A conservation law of a system can be defined as an equation in divergence free form, viz.,

$$
\operatorname{div} f = D_i f^i = 0, \qquad \text{(A3.3)}
$$

where f is a vector function $f\,(u, \underset{1}{u}, \underset{2}{u}, \dots, \underset{k}{u}) \equiv (f^1, f^2, \dots, f^n)$, and D_i ($\equiv D_{x_i}$) are the total derivatives to some order k. Eq. (A3.3) must hold for any external function $u(x)$ of (A3.2). The vector f is also called conserved flux (Noether current) since eq. (A3.3) implies that a net flow of f through any closed surface in the space x is zero. Noether [42] originally considered the transformations (A3.1) in the form (cf. Sects. 2.3.3 and 6.1.1) ($\in$ is an infinitesimal parameter)

$$
\begin{aligned}
x'_i &= x_i + \in \xi_i(x, u, \underset{1}{u}, \underset{2}{u}, \dots\dots, \underset{p}{u}) + 0(\in^2), \qquad \text{(A 3.4)}\\
u' &= u + \in \eta\,(x, u, \underset{1}{u}, \underset{2}{u}, \dots\dots, \underset{p}{u}) + 0(\in^2),
\end{aligned}
$$

which leave the action integral $J(u)$ invariant for arbitrary Ω, and established the relationship between infinitesimals ξ_i, η and the conserved flux f.

An infinitesimal characterization of the group G corresponding to (A3.1) is defined by the transformation operator [254, 256],

$$X = \xi_i \frac{\partial}{\partial x_i} + \eta^\alpha \frac{\partial}{\partial u^\alpha} + \zeta_i^\alpha \frac{\partial}{\partial u_i^\alpha} + \ldots + \zeta_{i_1..i_k}^\alpha \frac{\partial}{\partial u_{i_1..i_k}^\alpha} + \ldots , \qquad \text{(A3.5)}$$

where

$$\xi_i = (\partial g_i / \partial t)_{t=0};\ \eta^\alpha = (\partial \phi^\alpha / \partial t)_{t=0};\ \zeta_{i_1..i_k}^\alpha = (\partial \psi_{i_1..i_k}^\alpha / \partial t)_{t=o} ,$$
$$(k = 1,2,..).$$

Here, for the case of Lie-Backlund groups one also considers the transformation laws for the derivatives of u in the infinitesimal form as

$$u_{i_1..i_k}^{\prime\alpha} = u_{i_1..i_k}^\alpha + \in \zeta_{i_1..i_k}^\alpha (x, u, \underset{1}{u}, \underset{2}{u}, \ldots, \underset{p}{u}) + 0(\in^2),$$

where ξ's are expressed in terms of D_i's through [254, 256]

$$\zeta_i^\alpha = D_i(\eta^\alpha) - u_j^\alpha D_i(\xi_j) ,$$
$$\zeta_{i_1 i_2}^\alpha = D_{i_2}(\zeta_{i_1}^\alpha) - u_{i_1 j}^\alpha D_{i_2}(\xi_j) ,$$
$$\vdots$$
$$\zeta_{i_1..i_k}^\alpha = D_{i_k}(\zeta_{i_1..i_{k-1}}^\alpha) - u_{i_1..i_{k-1}j}^\alpha D_{i_k}(\xi_j) ,$$

with

$$D_i = \frac{\partial}{\partial x_i} + u_i^\alpha \frac{\partial}{\partial u^\alpha} + u_{ii_1}^\alpha \frac{\partial}{\partial u_{i_1}^\alpha} + u_{ii_1i_2}^\alpha \frac{\partial}{\partial u_{i_1i_2}^\alpha} + \ldots .$$

Thus, for any function F one has

$$F(x', u', \ldots) = F(x, u, \ldots.) + \in X F(x, u, \ldots) + O(\in^2).$$

The generator X is further used to look for the symmetries of the differential equation $\omega(x_i, u^\alpha, u_i^\alpha, u_{i_1 i_2}^\alpha, \ldots) = 0$ by demanding $X\omega = 0$ such that the differential consequences, namely $\omega = 0$, $D_i\omega = 0$, $D_{i_1}(D_{i_2}\omega) = 0, \ldots.$, are satisfied simultaneously. For details we refer to the recent literature [256, 257]. A study of Hamilton-Jacobi equation within this framework is carried out by Fokas [254]. As far as the determination of conservation laws or the corresponding constants of motion is concerned, they are the consequences of eq. (A3.3). In general this equation provides the invariants for the corresponding Lagrangian system, but for the systems in classical mechanics where t is the only independent variable, eq. (A3.3) yields $D_t f^1 = 0$ and f^1 becomes the constant of motion. Studies have also been carried out [257] for non-Lagrangian systems.

References

1. A. Khare and S.N. Behra, Pramana-J. Phys. **14** (1980) 327.
2. D. Amin, Phys. Today **35** (1982) 35; Phys. Rev. Lett. **36** (1976) 323.
3. S. Coleman, "*Aspects of Symmetry*" selected Erice Lectures (Univ. Press, Cambridge, 1988) p. 234.
4. See, for example, R. Damburg, R. Propin and S. Graffi, Int. J. Quant. Chem. **21** (1982) 191; 195.
5. See, for example, A. Ghatak and K. Thiagarajan, in Prog. in Optics ed. by E. Wolf (North Holland, 1980).
6. See, for example, the Aharonov-Bohm potential discussed in Ref. (135).
7. S.N. Behra and A. Khare, Lett. Math. Phys. **4** (1980) 153; Pramana-J. Phys. **15** (1980) 501.
8. See, for example, J.L. Powell and B. Crasemann, "*Quantum Mechanics*" (Addison-Wesley Pub. Co. Inc, 1961) p. 387.
9. E.T. Whittaker, "*Analytical Dynamics*" (Univ. Press, Cambridge, 1927).
10. B. Eckhardt, Phys. Rep. **163** (1988) 205-297.
11. H. Yoshida, Phys. Lett. **A120** (1987) 388; D. Roekaerts and H. Yoshida, J. Phys. **A21** (1988) 3547; H. Yoshida, Commun. Math. Phys. **116** (1988) 529.
12. J. Hietarinta, Phys. Rep. **147** (1987) 87-154.
13. C.R. Holt, J. Math. Phys. **23** (1982) 1037.
14. L.S. Hall, Physica **D8** (1983) 90.
15. N.N. Bogoliubov and Y.A. Mitropolski, "*Asymptotic Methods in the Theory of Nonlinear Oscillations*" (Gordon and Breach, New York, 1961); M.D. Kruskal, J. Math. Phys.**3** (1962) 806.
16. V.I. Arnold, "*Mathematical Methods of Classical Mechanics*" (Springer, New York, 1978).
17. R. Abraham and J.E. Marsden, "*Foundations of Mechanics*" 2nd Ed. (Benjamin/Cummings, Reading, 1982).
18. M. Lakshmanan and R. Sahadevan, Phys. Rep. **224** (1993) 1-93.
19. D.C. Khandekar and S.V. Lawande, Phys. Rep. **137** (1986) 115-229.
20. R.S. Kaushal, D. Parashar and S.C. Mishra, Fortschritte der Phys. (Berlin) **42** (1994) 689-705, No. 8
21. G. Darboux, Sur un Probleme de mechanique, Archires Neerlandaises (ii) **6** (1901) 371, as cited in Ref. (12) above.
22. B. Dorizzi, B. Grammaticos and A. Ramani, J. Math. Phys. **24** (1983) 2282; A. Ankiewicz and C. Pask, J. Phys. **A16** (1983) 4203; G. Thompson, J. Phys. **A17** (1984) 985; T. Sen, Phys. Lett. **A111** (1985) 97.

23. R.S. Kaushal, S.C. Mishra and K.C. Tripathy,, J. Math. Phys. **26** (1985) 420.
24. R.S. Kaushal, S.C. Mishra and K.C. Tripathy, Phys. Lett. **A102** (1984) 7.
25. S.C. Mishra, R.S. Kaushal and K.C. Tripathy, J. Math. Phys. **25** (1984) 2217.
26. S.C. Mishra Ph.D. Thesis "Some Studies on Two-Dimensional Classical Integrable Systems" (Unpublished) Delhi University, 1985.
27. R.S. Kaushal and S.C. Mishra, Pramana-J. Phys. **26** (1986) 109.
28. J. Hietarinta, Phys. Rev. Lett. **52** (1984) 1057.
29. H.R. Lewis and P.G.L. Leach, Ann. Phys. **164** (1985) 47; J. Goedert and H.R. Lewis, J. Math. Phys. **28** (1987) 728; 736.
30. S. Ichtiaroglou and G. Voyatzis, J. Phys. **A21** (1988) 3537.
31. J. Hietarinta, Phys. Lett. **A96** (1983) 273.
32. A.S. Fokas and P.A. Lagerstrom, Journ. Math. Anal. & Appl. **74** (1980) 325.
33. V.I. Inozemtsev, Phys. Lett **A96** (1983 447.
34. M. Toda, J. Phys. Soc. Jpn. **23** (1967) 501; S.P. Ford et al, Prog. Theo. Phys. **50** (1973) 1547; M. Toda, "*Theory of Nonlinear Lattices*" (Springer, Berlin, 1980).
35. B. Grammaticos, B. Dorizzi and A. Ramani, J. Math. Phys. **25** (1984) 3470.
36. P. Painleve' Bull. Soc. Math. **28** (1900) 201; Acta Math. **25** (1902) 1.
37. L. Fuchs, Sitz. Akad. Wiss. Berlin **32** (1884) 699; S. Kovalevskaya, Acta. Math. **12** (1889) 177; ibid. **14** (1889) 81.
38. B. Grammaticos, B. Dorizzi and A. Ramani, J. Math. Phys. **24** (1984) 2289; ibid. **25** (1984) 3470, A. Ramani, B. Grammaticos and T. Bountis, Phys. Rep. **180** (1989) 159; W.H. Steeb and N. Euler, "*Nonlinear Evolution Equations and Painleve' Test*" (World Scientific, 1988).
39. P.D. Lax Comm. Pure Appl. Math. **21** (1968) 467; for a recent review of this method see, for example, M.A. Olshanesky and A.M. Perelomov, Phys. Rep. **71** (1981) 313; and Steeb and Euler in Ref. (38) above.
40. F. Calogero, Lett. Nuovo Cim. **13** (1975) 411; P.P. Kulish, Theo. Math. Phys. **26** (1976) 132.
41. J. Moser, Adv. Math. **16** (1975) 197; F. Calogero, Lett. Nuovo. Cim. **13** (1975) 383.
42. E. Noether, Nachr. Ges. Wiss. Gottingen Math. Phys. KL. (1918) 235.
43. M. Lutzky, J. Phys. **A12** (1979) 973.
44. P.G.L. Leach, J. Math. Phys. **26** (1985) 77.
45. W. Sarlet, F. Mahomed and P.G.L. Leach, J. Phys. **A20** (1987) 277.
46. M. Lakshmanan and R. Sahadevan, J. Math. Phys. **32** (1991) 75; ibid. **24** (1983) 795, R. Sahadevan and M. Lakshmanan, J. Phys. **A19** (1986) L949.
47. M.F. Ranada, J. Math. Phys. **35** (1994) 1219.
48. M. Lutzky, Phys. Lett. **A72** (1979) 86; ibid **A75** (1979) 8; J. Phys. **A15** (1982) L87.
49. W. Sarlet and F. Cantrijn, SIAM Rev. 23 (1981) 467; Phys. Lett. **A88** (1982) 383; G. Caviglia, Int. Jour. Theo. Phys. **23** (1984) 461.
50. M. Henon and C. Heiles, Astron. J. **69** (1964) 73; P.K. Kaw, A. Sen and E.J. Valeo, Physica **D9** (1983) 96.
51. Y. Aizawa and N. Saito, J. Phys. Soc. Jpn. **32** (1972) 1636; A. Ramani, B. Dorizzi and B. Grammaticos, Phys. Rev. Lett. **49** (1982) 1539; Also, see Ref. (12).

52. D.V. Choodnosky and G.V. Choodnosky, Lett. Nuovo Cim. **22** (1978) 47; Also, See H. Grosse, Acta Phys. Austr. **52** (1980) 89; B. Grammaticos, A. Ramani and H. Yoshida, Phys. Lett. **A124** (1987) 65.
53. S. Wojciechowski, Physica Scripta **31** (1985) 433.
54. S. Chandrasekhar, "*Principles of stellar Dynamics*" (Dover. Publ., N.Y. 1942) Ch. 3.
55. Y.F. Chang, M. Tabor and J. Weiss, J. Math. Phys. **23** (1982) 531; T. Bountis and H. Segur, in "Mathematical methods in Hydrodynamics & Integrability in dynamical systems", A.I.P. Conf. Proc. No. 88, ed. by M. Tabor and Y.M. Treve, 1981.
56. J. Moser, in "Topics in Nonlinear dynamics", A.I.P. Conf. Proc. No. 46, ed. by S. Jorna 1978.
57. K.J. Whiteman, Rep. Prog. Phys. **40** (1977) 1033.
58. P.A.M. Dirac, Proc. Roy. Soc. **A246** (1958) 326; J.L. Synge, "Classical dynamics", in Handbuch der Phys. Bd. III/1 ed. by S. Flugge, Springer Verlag, 1960, p.143; W. Thirring, "*Classical dynamical systems*" (Springer Verlag, 1978) p. 88.
59. See, for example, P. Helander, M. Lisak and V.E. Semenov, Phys. Rev. Lett. **68** (1992) 3659, and the references therein.
60. L.D. Landau and E.M. Lifshitz, "*Mechanics*" (Pergamon, Oxford, 1976), 3rd ed; H. Goldstein, "*Classical Mechanics*" (Addison-Wesley Reading, MA, 1980), 2nd ed.
61. M. Kolsrud, Phys. Rev. **104** (1956) 1186.
62. M. Kruskal, J. Math. Phys. **3** (1962) 806.
63. H.R. Lewis Jr., J. Math. Phys. **9** (1968) 1976; H.R. Lewis Jr., Phys. Rev. Lett. **13** (1967) 510, 636.
64. H.R. Lewis and W.B. Riesenfeld, J. Math. Phys. **10** (1969) 1458; S.S. Mizrahi, Phys. Lett. **A138** (1989) 465.
65. M.S. Abdalla and R.K. Colegrave, Phys. Rev. **32A** (1985) 1958; R.K. Colegrave and M.A. Mannan, J. Math. Rhys. **29** (1988) 1580.
66. R.S. Kaushal and H.J. Korsch, J. Math. Phys. **22** (1981) 1904.
67. P.G.L. Leach, SIAM J. Appl. Math. **34** (1978) 496.
68. J.R. Ray and J.L. Reid, Phys. Rev. **A26** (1982) 1042.
69. P.G.L. Leach, J. Math. Phys. **22** (1981) 465; **20** (1979) 96.
70. P.G.L. Leach and S.D. Maharaj, J. Math. Phys. **33** (1992) 2023.
71. P.G.L. Leach, J. Math. Phys. **26** (1985) 2510; W. Sarlet and L.Y. Bahar, Int. J. Nonlin. Mech. **15** (1980) 133.
72. J.R. Ray and J.L. Reid, J. Math. Phys. **20** (1979) 2054.
73. R.S. Kaushal, Pramana-J. Phys. **24** (1985) 663.
74. V.P. Ermakov, Univ. Izv. Kiev Series III **9** (1880) 1.
75. J.R. Ray and J.L. Reid, Phys. Lett **A71** (1979) 317; ibid. **A74** (1979) 23.
76. J.L. Reid and J.R. Ray, J. Math. Phys. **23** (1982) 503; J. Phys. **A15** (1982) 2751.
77. H.J. Korsch, Phys. Lett. **A74** (1979) 294.
78. K. Takayama, Phys. Lett. **A88** (1982) 57.
79. M. Lutzky, Phys. Lett. **A68** (1978) 3; J. Phys. **A11** (1978) 249.

80. J.R. Ray, Phys. Rev. **A26** (1982) 729.
81. J.R. Burgan et al., Phys. Lett. **A74** (1979) 11; J.R. Burgan et al., J. Plasma Phys. **19** (1978) 135.
82. M.R. Feix, S. Bouquet and H.R. Lewis, Physica **D28** (1987) 80.
83. H.R. Lewis and P.G.L. Leach, J. Math. Phys. **23** (1982) 2371; B. Grammaticos and B. Dorizzi, J. Math. Phys. **25** (1984) 2194.
84. G.H. Katzin and J. Levine, J. Math. Phys. **24** (1983) 1761; Also see, ibid. **18** (1977) 1267; **23** (1982) 552.
85. R.S. Kaushal and S.C. Mishra, J. Math. Phys. **34** (1993) 5843; R.S. Kaushal D. Parashar, Shalini Gupta and S.C. Mishra, Ann. Phys. (NY) **259** (1997) 233. (Referred to as KPGM in the text).
86. R.S. Kaushal, "Third order invariants for time dependent two-dimensional classical dynamical systems', unpublished.
87. P.G.L. Leach, Phys. Lett. **A158** (1991) 102; K.S. Govinder and P.G.L. Leach, Phys. Lett. **A186** (1994) 391; K.S. Govinder and P.G.L. Leach, J. Phys. **A27** (1994) 4153.
88. C. Athorne, Phys. Lett. **A159** (1991) 375.
89. E. Pinney, Proc. Am. Math. Soc. **1** (1950) 681.
90. C. Athorne, J. Phys. **A24** (1991) 945; K.S. Govinder, C. Athorne and P.G.L. Leach, J. Phys. **A26** (1993) 4035.
91. J.M. Cervero and J.D. Lejarreta, Phys. Lett. **A156** (1991) 201.
92. R.S. Kaushal, Pramana-J. Phys. **42** (1994) 467.
93. A.M. Perelomov, "*Integrable Systems of Classical Mechanics and Lie Algebras*" Vol. 1. (Birkhauser Verlag, 1990).
94. R.K. Colegrave, P. Croxson and M.A. Mannan, Phys. Lett. **A131** (1988) 407.
95. H.R. Lewis and P.G.L. Leach, J. Math. Phys. **23** (1982) 165.
96. See, for example, A. Ghatak and K. Thiagrajan in Ref. (5) and the Refs. (113) and (122) below; U. Jain et al. J. Opt. Soc. Am. **72** (1982) 1545, R.S. Kaushal, J. Opt. Soc. Am. **A8** (1991) 1245.
97. R.S. Kaushal, Ann. Phys. **206** (1991) 90.
98. P.M. Mathews, M. Seetharaman and S. Raghavan, J. Phys. **A14** (1982) 103; G.P. Flessas, Phys. Lett. **A95** (1983) 361; K. Banerjee and J.K. Bhattacharjee, Phys. Rev. **D29** (1984) 1111.
99. See, for example, L.I. Schiff, "*Quantum Mechanics*", 3rd Ed. (McGraw-Hill, New York, 1968) p. 81.
100. R.S. Kaushal, Pramana -J. Phys. **42** (1994) 315.
101. G.P. Flessas, Phys. Lett. **A72** (1979) 289; G.P. Flessas and K.P. Das, Phys. Lett. **A78** (1980) 19.
102. M. Znojil, J. Phys. **A15** (1982) 2111.
103. G. Levai, J. Phys. **A22** (1989) 689; **A25** (1992) L521; **A27** (1994) 3809.
104. A.P. Hautot, Phys. Lett. **A38** (1972) 305; E. Magyari, Phys. Lett. **A81** (1981) 116.
105. A.K. Dutta and R.S. Wiley, J. Math. Phys. **29** (1988) 893.
106. P. Roy and R. Roy Chaudhury, Phys. Lett. **A122** (1987) 275.
107. A.V. Turbiner and A.G. Ushveridze, Phys. Lett. **A126** (1987) 181.
108. A.V. Turbiner, Commun. Math. Phys. **118** (1988) 467.

109. V. Singh, S.N. Biswas and K. Datta, Phys. Rev. **D18** (1978) 1901.
110. A. de Souza Dutra, Phys. Lett. **A131** (1988) 319.
111. R.S. Kaushal, Phys. Lett. **A142** (1989) 57.
112. R.S. Kaushal, Mod. Phys. Lett. **A6** (1991) 383.
113. H.H. Aly and A.O. Barut, Phys. Lett. **A145** (1990) 299, and the references therein.
114. L.A. Singh, S.P. Singh and K.D. Singh, Phys. Lett. **A148** (1990) 389.
115. R.S. Kaushal and D. Parashar, Phys. Lett. **A170** (1992) 335.
116. M. Landtman, Phys. Lett. **A175** (1993) 147.
117. Y.P. Varshni, Phys. Lett. **A183** (1993) 9.
118. M. Znojil and P.G.L. Leach, J. Math. Phys. **33** (1992) 2785.
119. R. Adhikari, R.Dutt and Y.P. Varshni, Phys. Lett. **A141** (1989) 1.
120. R. Adhikari, R. Dutt and Y.P. Varshni, J. Math. Phys. **32** (1991) 447.
121. F. Cooper, A. Khare and U. Sukhatme, Phys. Rep. **251** (1995) 267.
122. M. Simsek and S. Ozcelik, Phys. Lett. **A186** (1994) 35; and the references therein.
123. A. de Souza Dutra and H. Boschi Filho, Phys. Rev. **A44** (1991) 4721.
124. B.L. Burrows, M. Cohen and T. Feldman, J. Math. Phys. **35** (1994) 5572.
125. M.F. Manning, Phys. Rev. **48** (1935) 161; A. Bhattacharji and E.C.G. Sudarshan, Nuovo Cim. **25** (1962) 864; A.K. Bose, Nuovo Cim. **32** (1964) 679.
126. S.K. Bose and N. Varma, Phys. Lett. **A141** (1989) 141.
127. I.S. Gradshteyn and I.M. Ryzhik, "*Table of integrals, series and products*" (Academic Press, New York, 1969) p. 337
128. E. Witten, Nucl. Phys. **B185** (1981) 513; F. Cooper and B. Freedman, Ann. Phys. **146** (1983) 262.
129. M.M. Nieto, Phys. Lett. **B145** (1984) 208.
130. J. Beckers, N. Debergh and L. Ntibashirakandi, J. Math, Phys. **36** (1995) 641; A. Kostelecky and M.M. Nieto, Phys. Rev. Lett. **53** (1984) 2285.
131. A. Khare and R.K. Bhaduri, Am. J. Phys. **62** (1994) 1008; A. Khare and R.K. Bhaduri, J. Phys. **A27** (1994) 2213.
132. F. Calogero, J. Math. Phys. **10** (1969) 2191.
133. D.R. Taylor and P.G.L. Leach, J. Math. Phys. **30** (1989) 1525.
134. J. Makarewicz, J. Phys. **A16** (1983) L553.
135. Ashok Guha and S. Mukherjee, J. Math. Phys. **28** (1987) 840.
136. A.V. Turbiner and A.G. Ushveridze, "On exact solutions of multidimensional Schrodinger equation with polynomial potentials", ITEP Preprint No. 86-169, Pub. in Kratkie Soobtschenia Fizika **12** (1988) 37-41.
137. R.S. Kaushal, Indian J. Phys. **68B** (1994) 409.
138. A.J. Markworth, Am. J. Phys. **45** (1977) 640.
139. F.T. Hioe, D. MacMillan and E.W. Montroll, Phys. Rep. **43** (1978) 305.
140. See, for example, Refs. (1-3) and L.A. Carreira in "Topics in current chemistry" ed. F.L. Boschke (Springer Verlag, Berlin, 1979) Ch.1.
141. M. Tabor, "*Chaos and Integrability in Nonlinear Dynamics*" (John Wiley & Sons, NY, 1989) p. 236; p. 70.

142a. M.R.M. Witwit, J. Math. Phys. **33** (1992) 4196; M.R.M. Witwit, J. Phys.

A24 (1991) 4535; M.R.M. Witwit and J.P. Killingbeck, Pramana-J. Phys. **43** (1994) 279.
142b. M.R.M. Witwit, J. Math. Phys. **34** (1993) 5050; ibid.**33** (1992) 2779.
143. R.K. Agrawal and V.S. Varma, Phys. Rev. **A49** (1994) 5089; ibid. **A48** (1993) 1921.
144. R. Guardiola and J. Ros, J. Phys. **A25** (1992) 1351.
145. C.M. Bender and A.V. Turbiner, Phys. Lett. **A173** (1993) 442.
146. A.de Souza Dutra, A.S. de Castro and H. Boschi-Filho, Phys. Rev. **A51** (1995) 3480.
147. P. Winternitz, J.A. Somorodinsky, M. Uhlir and I. Fris, Sov. J. Nucl. Phys. **4** (1967) 444; A.A. Makarov, J.A. Somorodinsky, Kh. Valiev and P. Winternitz, Nuovo Cim. **A52** (1967) 1061.
148. E. Wichmann, J. Math. Phys. **2** (1961) 876.
149. J. Wei and E. Norman, J. Math. Phys. **4** (1963) 575; Proc. Am. Math. Soc. **15** (1963) 327.
150. F. Wolf and H.J. Korsch, Phys. Rev. **A37** (1988) 1934.
151. F.M. Fernandez, J. Math. Phys. **30** (1989) 1522.
152. F.M. Fernandez, J. Phys. **A21** (1988) 1357; F.M. Fernandez and E.A. Castro, Phys. Lett. **A125** (1987) 77.
153. F. Salmistraro and R. Rosso, J. Math. Phys. **34** (1993) 3964.
154. J.G. Hartley and J.R. Ray, Phys. Rev. **A24** (1981) 2873.
155. J.G. Hartley and J.R. Ray, Phys. Rev. **A25** (1982) 2388; J.R. Ray and J.G. Hartley, Phys. Lett. **A88** (1982) 125.
156. D.C. Khandekar and S.V. Lawande, J. Math. Phys. **16** (1975) 1458.
157. P.G.L. Leach, J. Math Phys. **18** (1977) 1608; 1902.
158. P. Camiz et. al., J. Math. Phys. **12** (1971) 2040.
159. D.C. Khandekar and S.V. Lawande, J. Math. Phys. **20** (1979) 1870.
160. D.R. Truax, J. Math. Phys. **23** (1982) 43; ibid. **22** (1981) 1959.
161. R.P. Feynman, Rev. Mod. Phys. **20** (1948) 367; Phys. Rev. **84** (1951) 108.
162. R.P. Feynman and A.R. Hibbs, "*Quantum Mechanics and Path Integrals*" (McGraw-Hill, New York, 1965) p. 58
163. J.H. Van Vleck, Proc. Nat. Acad. Sci. (USA) **14** (1928) 178; A.V. Jones and G.J. Papadopoulos, J. Phys. **A4** (1971) L86.
164. M.J. Goovaerts, Physica **77** (1974) 379; G.J. Papadopoulos, Phys. Rev. **D11** (1975) 2870.
165. M. Kleber, Phys. Rep. **236** (1994) 331-393; H.P. Breur and M. Holthaus, Ann. Phys. (NY) **211** (1991) 249; F. Bensch, H.J. Korsch, B. Mirbach and N. Ben-Tal, J. Phys. **A25** (1992) 6761; B. Mirbach and H.J. Korsch, J. Phys. **A27** (1994) 6579.
166. R.S. Kaushal, Phys. Rev. **A46** (1992) 2941.
167a. S.C. Mishra and D. Parashar, Int. Journ. Theo. Phys. **29** (1990) 299.
167b. M. Lakshmanan and R. Sahadevan, Phys. Rev. **A31** (1985) 861; R. Sahadevan and M. Lakshmanan, Phys. Rev. **A33** (1986) 3563; ibid. **A34** (1986) 5137.
168. S. Datta Majumdar and M.J. Englefield, Int. J. Theo. Phys. **16** (1977) 829.
169. C.P. Boyer, Hel. Phys. Acta **47** (1974) 589; C.P. Boyer, R.T. Sharpand, P. Winternitz, J. Math. Phys. **17** (1976) 1439.

170. C.J. Eliezer and A. Gray, SIAM J. Appl. Math. **30** (1976) 463.
171. R.S. Kaushal and D. Parashar, "Quantum mechanics and supersymmetric quantum mechanics as the multi-dimensional Ermakov theories", Proc. XI DAE Symp., Santiniketan (India), Dec. 28, 1994-Jan.2, 1995, vol.1, p.161; J. Phys. **A29** (1996) 889; Also, see R.S. Kaushal, in Proc. IV Int. Wigner Symp. ed. by T.H. Saligman (World Scientific Pub. Co., Singapore, 1996) p. 131.
172. H.J. Korsch and H. Laurent, J. Phys. **B14** (1981) 4213; Also see, H.J. Korsch, Phys. Lett. **A109** (1985) 313.
173. H.J. Korsch, H. Laurent and R. Mohlenkamp, J. Phys. **B15** (1982) 1.
174. R.A. Lee, J. Phys. **A17** (1984) 535; ibid. **A15** (1982) 2761.
175. D.C. Khandekar and S.V. Lawande, Phys. Lett. **A67** (1978) 175; D.C. Khandekar and S.V. Lawande, J. Math. Phys. **16** (1975) 384; S.V. Lawande and A.K. Dhara, Phys. Lett. **A99** (1983) 353.
176. A.K. Dhara and S.V. Lawande, J. Phys. **A17** (1984) 2423; D.C. Khandekar and S.V. Lawande, J. Phys. **A5** (1972) 612.
177. S. Chandrasekhar, *"An Introduction to the study of Steller Structure"* (Dover Publications, Inc, Univ. of Chicago, 1939) Chap. 3, 4.
178. B.K. Berger, Phys. Rev. **D18** (1978) 4367.
179. C.W. Misner, Phys. Rev. **D8** (1973) 3271.
180a. J.R. Ray, Phys. Rev. **D20** (1979) 2632.
180b. R. S. Kaushal, Clas, & Quantum Grav. **14** (1997) 1.
181. J. Martin and S. Bouquet, J. Math. Phys. **35** (1994) 181.
182. H.J. Giacomini, J. Phys. **A23** (1990) 587; 865.
183. J.L. Reid, Proc. Am. Math. Soc. **38** (1973) 532.
184. J.M. Thomas, Proc. Am. Math. Soc. **7** (1956) 95.
185. P.B. Burt and J.L. Reid, J. Math. Analys. & Appl. **55** (1976) 43.
186. A.L. Hodgkin and A.F. Huxley, J. Physiol. **116** (1952) 449; ibid. **117** (1952) 500; R. Fitz Hugh, Biophys. J. **1** (1961) 445, For other earlier works see, for example, A.C. Scott. Rev. Mod. Phys. **47** (1975) 505.
187. J.L. Hindmarsh and R.M. Rose, Nature **296** (1982) 162.
188. M. Lakshamanan and K. Rajagopal, Phys. Lett. **A82** (1981) 266.
189. K. Rajagopal, Phys. Lett.. **A98** (1983) 77; ibid. **A99** (1983) 261; ibid. **A100** (1984) 49; ibid. **A105** (1984) 160; ibid. **A108** (1985) 228.
190. H.R. Lewis, "Representation of magnetic fields with toroidal topology in terms of field-line invariants", Report No. LA-UR-88-2607 Revised, Los Alamos National Lab.; Proc. Int. Conf. Plasma Phys., Delhi (India) 1989, paper H-19.
191. H.R. Lewis and B. Abraham-Shrauner, Bull. Am. Phys. Soc, **34** (1989) 1974.
192. A.H. Boozer, Phys. Fluids **26** (1983) 1288.
193. G.K. Savvidy, Phys. Lett. **B130** (1983) 303; G.K. Savvidy, Nucl. Phys. B246 (1984) 302; S.J. Chang, Phys. Rev. **D29** (1984) 259.
194. J. Villarroel, J. Math. Phys. **29** (1988) 2132.
195. C. Nagaraj Kumar and A. Khare, Preprint, Inst. of. Phys., Bhubneswar, 1987; C. Nagaraj Kumar and A. Khare, J. Phys. **A22** (1989) L849.
196. R.J. Glauber, in *Quantum Optics and Electronics*, ed. by C.D.A. Blandin

and C. Cohen-Tannoudji (Gordon and Breach, New York, 1965); Phys. Rev. **131** (1963) 2766; Also see special issues of J. Mod. Opt. **34** (1987), J. Opt. Soc. Am. **B4** (1987); For more technical details see A. Perelomov, "*Generalized Coherent States and their Applications*" (Springer Verlag, 1986).

197. V.V. Dodonov and V.I. Man' ko, in "*Invariants and Evolution of Nonstationary Quantum Systems*" ed. by M.A. Markov, Proc. Lebedev Phys. Inst. Vol. 183 (Nova Science, Commack, N.Y. 1989).
198. I.A. Malkin and V.I. Man' ko, Phys. Lett. **A32** (1970) 243; I.A. Malkin, V.I. Man' ko and D.A. Trifonov, Phys. Rev. **D2** (1970) 1371.
199. V.V. Dodonov, E.V. Kurmyashev and V.I. Man' ko, Phys. Lett. **A79** (1980) 150.
200. K. Husimi, Prog. Theo. Phys. **9** (1953) 381.
201. V.I. Man' ko, in Proc. Group Theory Conf. Osaka (Japan) ed. by Arima (World Scientific, 1995), p. 320; Also, see in Proc. of A.I.P. Lectures given at Latin Am. School of Physics (1995) and in Proc. Fourth Int. Wigner Symp. held at Guadalajara (Mexico) during Aug. 7-11, 1995; R.S. Kaushal and D. Parashar, Phys. Rev. **A55** (1997) 2610.
202. See, for example, O. Castanos, R. Lopez-Pena and V.I. Man' ko, Phys. Rev. **A50** (1994) 5209 and the references therein.
203. M.V. Berry, Proc. R. Soc. **A392** (1984) 45.
204. B. Simon, Phys. Rev. Lett.. **51** (1983) 2167; D. Arovas, J.R. Schrieffer and F. Wilczek, Phys. Rev. Lett. **53** (1984) 722; G.W. Semenoff and P. Sodano, Phys. Rev. **57** (1986) 1195.
205. F.D.M. Haldane and Y.S. Wu, Phys. Rev. Lett. **55** (1985) 2887.
206. M. Stone, Phys. **D33** (1986) 1191; J. Moody, A. Shapere and F. Wilczeck, Phys. Rev. Lett. **56** (1986) 893; R. Jackiw, Phys. Rev. Lett. **56** (1986) 2779; J.C. Martinez, Phys. Lett. **A127** (1988) 399.
207. J.H. Hannay, J. Phys. **A18** (1985) 221; E. Gozzi and W.D. Thacker, Phys. Rev. **D35** (1987) 2388; 2398; G. Ghosh and B. Dutta-Roy, Phys. Rev. **D37** (1988) 1709; M. Kugler and S. Strikman, Phys. Rev. **D37** (1988) 934.
208. F. Wilczek and A. Zee, Phys. Rev. Lett. **52** (1984) 2111; H.Z. Li, Phys. Rev. Lett. **58** (1987) 539; J. Samuel and R. Bhandari, Phys. Rev. Lett. **60** (1988) 2339; J.Q. Liang and H.J.W. Muller-Kirsten, Ann. Phys. (NY) **219** (1992) 42.
209. A. Shapere and F. Wilczek, eds. "*Geometric Phases in Physics*" Ad. Series in Math. Phys. Vol. 5 (World Scientific, Singapore, 1989).
210. C.M. Cheng and P.C.W. Fung, J. Phys. **A22** (1989) 3493.
211. A. Aharonov and J. Anandan, Phys. Rev. Lett. **58** (1987) 1593.
212. J. Anandan, Phys. Lett. **A133** (1988) 171.
213. M.V. Berry, Proc. R. Soc. **A414** (1987) 31-46.
214. J.C. Garrison and R.Y. Chiao, Phys. Rev. Lett. **60** (1988) 165.
215. D.J. Moore and G.E. Stedman, J. Phys. **A23** (1990) 2049.
216. D.B. Monteoliva, H.J. Korsch and J.A. Nunez, J. Phys. **A27** (1994) 6897.
217. A.H. Zewail, "*Femtochemistry—Ultrafast Dynamics of Chemical Bonds*" Vol. I and II (World Scientific, Singapore, 1994).
218. E.C.G. Sudarshan and N. Mukunda, "*Classical Dynamics: A Modern Perspective*" (John Wiley & Sons N.Y., 1974).

219. P.A.M. Dirac, "*Lectures on Quantum Mechanics*" (Belfer Graduate School of Science Pub., Yeshiva Univ., New York, 1964) Chaps 1, 2.
220. P.A.M. Dirac, Can. Journ. Math. **2** (1950) 129; Proc. Roy. Soc. **A246** (1958) 326.
221. See, for example, X. Gracia and J.M. Pons, Ann. Phys. (NY) **187** (1988) 368; J. Barcelos-Neto, Ashok Das and W. Sherer, Acta Phys. Polonica **B18** (1987) 269; Proc. of XXVI Cracow School of Theo. Phys. Zakopane, Poland, 1986.
222. H.J.W. Muller-Kirsten, "Dynamical Systems with Constraints" Lecture Course, Summer 1994, Univ. of Kaiserslautern.
223. K. Sundermeyer, "*Constrained Dynamics*", Springer Lectures, Vol. 169 (Springer-Verlag, New York, 1982).
224. R. Vilela Mendes, "Topics in the quantum theory of non-integrable systems" Lectures at XXVII Karpacz Winter School of Theo. Phys., Feb. 1991, CERN Preprint No. CERN-TH-6034/91, March 1991.
225. See, for example, J. Howland, Ann. Inst. Henri Poincare **A50** (1989) 309; 325.
226. K. Yajima, J. Math. Soc. Jpn. **29** (1977) 729.
227. F. Bensch, H.J. Korsch, B. Mirbach and N. Ben-Tal, J. Phys. **A25** (1992) 6761; B. Mirbach and H.J. Korsch, J. Phys. **A27** (1994) 6579.
228. N. Ben-Tal, N. Moisyev and H.J. Korsch, Phys. Rev. **A46** (1992) 1669; N. Ben-Tal et al., Phys. Rev. **E47** (1993) 1646.
229. B. Mirbach and H.J. Korsch, Phys. Rev. Lett. **75** (1995) 362; N. Moiseyev, H.J. Korsch and B. Mirbach, Z. Physik **D29** (1994) 125.
230. H.P. Breuer and M. Holthaus, Ann. Phys. (NY) **211** (1991) 249.
231. M.C. Gutzwiller, J. Math. Phys. **12** (1971) 343; M.C. Gutzwiller, "*Chaos in Classical and Quantum Mechanics*" (Springer Verlag, 1990)
232. F.M. Izrailev, Phys. Rep. **196** (1990) 299; Also, see Lectures in "Miniworkshop on nonlinearity" (I.C.T.P) July 26-Aug.6, 1993.
233. B.D. Simons, P.A. Lee and B.L. Altschuler, Phys. Rev. Lett. **70** (1993) 4122; B. Southerland, J. Math. Phys. **12** (1971) 246; 251.
234. F.J. Dyson, J. Math. Phys. **3** (1962) 140.
235. P.G. Harper, Proc. Roy. Soc. London **A68** (1955) 874; T. Geisel, R. Ketzmerick and G. Petschel, Phys. Rev. Lett. **66** (1991) 1651.
236. T. Geisel, R. Ketzmerich and G. Petschel, Phys. Rev. Lett. **67** (1991) 3635, Also, see, T. Geisel et al., in "*Quantum Chaos: Between Order and Disorder*" ed. by G. Casati and B.V. Chirikov, Cambridge Univ. Press.
237. R. Lima and D.L. Shepelyansky, Phys. Rev. Lett. **67** (1991) 1377; P. Leboenf et al., Phys. Rev. Lett. **65** (1990) 3076.
238. J. Ambjorn and P. Olesen, Nucl. Phys. **B170** (1980) 60; 265.
239. H. Arodoz, Phys. Rev. **D27** (1983) 1903.
240. E.S. Nikolaevskii and L.N. Shur, JETP Lett. **36** (1982) 218.
241. G.Z. Baseyan, S.G. Mantinyan and G.K. Savvidy, JETP Lett. **29** (1979) 587; H.M. Asatryan and G.K. Savvidy, Phys. Lett. **A99** (1983) 290.
242. A. Carnegie and I.C. Percival, J. Phys. **A17** (1984) 801; S. Ichtiaroglou, J. Phys. **A22** (1989) 3461.

243. P. Dahlquist and G. Russberg, Phys. Rev. Lett. **65** (1990) 2837.
244. J. Shewell, Am. J. Phys. **27** (1959) 16; G.S. Agarwal and E. Wolf, Phys. Rev. **D2** (1970) 2161; L. Lohen, J. Math. Phys. **17** (1976) 597.
245. J. Moyal, Proc. Cambridge Philos. Soc. **45** (1949) 99.
246. See, for example, Refs. (10) and (224).
247. Susumu Okubo and Ashok Das, Phys. Lett. **B209** (1988) 311.
248. H.J. Korsch, Phys. Lett. **A90** (1982) 113.
249. J. Hietarinta, J. Math. Phys. **25** (1984) 1833; J. Hietarinta, Phys. Rev. **D25** (1982) 2103; Phys. Lett. **A93** (1982) 55.
250. J. Hietarinta and B. Grammaticos, J. Phys. **A22** (1989) 1315.
251. A.S. Fokas and P.A. Lagerstrom, J. Math. Anal. & Appl. **74** (1980) 342.
252. S. Kowalevskaya, Acta Math. **14** (1889) 81.
253. A. Ramani, B. Grammaticos and B. Dorizzi, Phys. Lett. **A101** (1984) 69.
254. A.S. Fokas, J. Math. Anal. & Appl. **68** (1979) 347.
255. N.H. Ibragimov and R.L. Anderson, J. Math. Anal. & Appl. **59** (1977) 145.
256. For the details of Lie-Backlund groups see, for example, G.W. Bluman and S. Kumei, "*Symmetries and Differential Equations*" (Springer Verlag, 1989) p. 252; P.J. Olver, "*Applications of Lie Groups to Differential Equations*" (Springer Verlag, 1986).
257. V. Rosenhaus and G.H. Katzin, J. Math. Phys. **35** (1994) 1998, and the references therein.

Index